CRYSTALLINE COMMUNION 2000

Mineral Properties
for Healing and Integration

by

Collen Marquist
and
Jack Frasl

This information was gained by
study of the inner and outer worlds.

Reprinted in Hong Kong in 2004

Earthlight, Inc.
46 Lakeshore Plaza
Kirkland, WA 98033
jackfrasl@yahoo.com

ISBN 0-9620201-4-1

Cover Design, Editing and Layout By: Pam Younghans, NorthPoint Graphics

You may order copies by calling
1-425-828-3872

TABLE OF CONTENTS

Introduction

Rocks, more specifically minerals, are tools. As humans we understand our dependence on the mineral kingdom, because all of our historical eras and advancements are named after rocks: the Stone Age, Bronze Age, Iron Age. We call our present era the Information Age, but underlying this stream of data packets is silicon or quartz, depending on which you choose to focus, and the more appropriate name will likely be applied in the future after we have fully grasped the signficance of the time we're living now.

Energy emanating from rocks is like light emitted from an incandescent light bulb. It is non-specific. The energy released from minerals is like lasers; specific, highly organized and potent.

This book is about the healing and consciousness integration properties of minerals. We say very little about the massive amount of country rock, fossils and shells which make up our planet. Each region of the world is defined by its underlying geology. People, animals, and plants evolve in concert with the directives given by local rocks.

But we ignored that discussion to focus on minerals as discreet sources of light, organized bundles of molecular signature spin, and storehouses of information. Minerals are light sources which emit discreet increments of information based on color, chemistry and structure. All minerals do not do all things. Minerals are the original kingdom of diversity. They are more individual than even plants and animals. They are the antithesis of "undifferentiated whole."

We may not have captured what is unique and holy about each mineral described here, but we've made every effort to notice what is authentically mineral and keep ourselves out of the vibration striving to characterize itself. You'll be the judge of accuracy based on your own experience. By all means, to thine own self be true.

We have noticed that the traditional light centers or chakras of the body relate to minerals based on chakra color, gland chemistry and polarity, the most important factors among many. Striving to form bridges of understanding and communion between humans and minerals, we have organized this book alphabetically by mineral, and list chakra associations in the individual readings as well as later in the book in an appendix which lists minerals under chakras. Chemistry is offered immediately, because the elements, combined in unique structures, determine physical and subtle body effects.

In ages past, local toes sunk into local sand and mud did the work of consciousness growth, but global mobility is the hallmark of our age. We encourage you to let your fingers do the walking for minerals. This era of mobility allows minerals to do more specific work on a global level than before achieved in human history. An experiment is reaching its pinnacle of exponential and ecstatic possibility through the alliance of carbon-based life forms and silicon and silicate minerals. We've made these connections before in Atlantis and England

and Egypt, to name a few. Despite some failures, we've managed to survive ourselves. The intention of this book is to assist humans to thrive in concert with the needs of Earth.

This book compiles information originally published in three smaller volumes as *Crystalline Communion Volume 1* (1987), *Volume 2* (1991), and *Volume 3* (1995). The mineral entries number approximately 1,200, and each has been updated and edited for accuracy based on our ongoing experiences working with minerals, people, the land, plants and animals. Since 1995, we've written a monthly column, "Messages in Stone," on mineral properties for the Seattle-based metaphysical newspaper *The New Times*. This column has been an invaluable resource for us to observe patterns in these readings which may have taken much longer to pinpoint. These patterns will be more fully described in our next book about the elements, as in the Periodic Table of the Elements.

We have learned through our work with rocks and minerals that consciousness is an evolutionary process, and occurs on the largest scale according to processes natural to our universe. Rocks and minerals persist in large time cycles, and they show us how long the consciousness cycle has taken to evolve.

Mineral healing and consciousness information is being discovered and rediscovered globally, and particularly by Western culture, which indicates to us that humans are opening to the feminine principle and the purple ray. Mineral frequencies act on people in clear and unbiased ways to promote balance, which is the natural order of things. Balance leads to health and the increased probability of spiritual growth and evolution of consciousness. Specific mineral vibrations assist with specific emotional, mental, physical and spiritual as well as subtle body goals and needs.

Having said that, let us make clear that the information offered here is not allopathic or in any way presumed to replace allopathic authority regarding medicine and traditional health care.

Instead, we are noting that all is vibration. Light and sound are indicators of motion and frequency. There is a sound, a tonal, a totem frequency to each individual which, when bonded with a mineral frequency on this planet, allows transformation, shape shifting, and light shifting or polarization.

During smaller increments of time, we assist wellbeing by harmonizing with universal energy patterns built into nature. The mineral kingdom is particularly useful to tune with due to the steady notes and frequencies its members produce. Their purity, constancy, reliability and patience are fine role models and supports for us in a chaotic time when we need both. This descriptive listing of crystal and mineral properties is one tool to assist the body's natural process of healing and growth.

In Appendix 1, a cross-reference by specific body zones or chakras lists minerals which impact each chakra. Each mineral listing following this introduction notes chakra associations in parentheses following the description. These are universal qualities which our unique needs modify from time to time. You will note that not all minerals have a specific enough area of influence to be placed exactly by chakra. Trust your intuitive powers to determine the specific impact of any mineral for you. Hold the stone, look at it, include it in your sphere of sensory influence, and see where and how you feel an effect. These personally tested responses are the most true and immediate synchronizing values for your being.

A Spiritual Apothecary in Appendix 2 lists mineral which, in particular, lend themselves to specific energetic and vibrational healing processes. This listing applies minerals to the auric, etheric and physical bodies.

Lastly, we encourage you to enjoy working, playing and living with minerals. They embody the most primal forces of the universe, atomic structure and the elements. Mineral influence is consistent, but because minerals are the foundation of consciousness, it takes focus and attention to notice their subtle energies as distinct from our own. This is because we are formed by the land on which we live. Being with mineral variety and beauty is enough to renew the spirit. It is our deepest hope that you will resonate and find your own truth within.

Mineral Properties for Healing and Integration

ABALONE: Abalone is a calcium carbonate seashell. Abalone is anticarcinogenic. It improves protein assimilation, and strengthens heart and other muscle tissue. The various colors of rainbow sheen enliven and impact many body areas when worn or viewed. It is a feminine, shamanic healing shell, and is the classic bowl used during smudge ceremony, a Native American cleansing ritual performed before sacred activity and to cleanse the self and areas on a regular basis. Abalone is the main food of the California sea otter, and embodies the healer energies of otter and the ocean. Abalone is a natural shapeshifting tool. It is highly regarded as a tool in the medicine bundle. *(All chakras)*

ACANTHITE: Acanthite is a green color, silver sulfide mineral in simple molecular arrangement forming in cubic building blocks. Acanthite is a good infection fighter (particularly streptococcal), and aids in counteracting all bacterial issues, as well as fungal and mold. It is excellent to use to counteract the effects of antibiotics. It may ameliorate anemia, diabetes, slow growth, and dermatitis. It aids the use of vitamin B1 and the development of hemoglobin, and helps cleanse the bloodstream and digestive tract. Large quantities of acanthite are difficult to locate normally, so the vibration quotient of acanthite can be increased by placing it with chlorite-included quartz in a healing layout, pouch or near the bed. Other minerals can be utilized depending on personal preference especially clear danburite, green tourmaline, aquamarine and emerald. To influence a room environment, place acanthite on a quartz cluster. It supports the healing properties of colloidal silver and sulfa medications. Its disinfectant abilities can be applied to needs on all physical, emotional, mental and subtle body levels. Ask for guidance through muscle testing, especially. *(1st and 5th chakras - lymph)*

ACTINOLITE: Actinolite is a hydrous calcium magnesium iron silicate with a green, elongated, diamond-shaped blade often found in mica or talc deposits. The tougher form of actinolite is known as nephrite jade. Actinolite raises the vibratory rate of the entire being as it removes attached negativities from the auric body, which then increases the receptivity of all the subtle and physical bodies to inspiration and healing energies. It is especially soothing for the heart chakra in conjunction with jade (which is primarily actinolite) family minerals. It acts as a teaching device during meditation, allowing integration of new awareness gained. Actinolite forms in a chemical series with tremolite (a form of jade) and ferroactinolite, and is the basic constituent of nephrite jade. Refer to grief and mourning minerals listed in the Index. It is a useful mineral for cleansing people suffering from recent loss of a loved one through death. Actinolite and all forms of jade in the amphibole group resonate to frog energy. Actinolite can be substituted for nephrite jade in the Auric Cleanser category of the Spiritual Apothecary. *(4th chakra)*

ADAMITE: Adamite is a hydrous zinc arsenate which can vary in color: colorless, yellow, green, pink, and violet.

Cobaltian (Purple) - "Ascension stone." This stone is particularly conducive to higher states of awareness. It opens the head region both physically and spiritually, so it is good to counteract sinus infections, colds and flu as well as being good for meditation, psychic and hypnotic work, and creative visualization. It supports prayer. *(7th chakra)*

Cuprian (Green) - This is a potent protection mineral, especially for people who shun or avoid modern technology such as cars and other moving objects. It helps increase a sense of being centered within a world so busy as to resemble utter chaos. Some people experience disorientation in cities, and this mineral provides calm and focus. *(4th chakra)*

Yellow - Adamite should always be near water, especially waterfalls, which they love. When placed in the left hand, adamite pushes one out of the body, quite forcefully at times. When held in the right hand, adamite causes energy flow to maximize. This mineral is more androgynous around water, and has strong polarizing qualities when used in a balanced situation such as around water rather than fire. *(3rd chakra)*

ADULARIA: Adularia is a translucent or transparent form of orthoclase which is a potassium aluminum silicate. It can be called a "glassy" feldspar. It is often found associated with quartz. Very nice specimens of adularia with smokey quartz come from Switzerland. Adularia is an excellent whitish mineral for healing skin disorders such as mild rash to psoriasis. Its character resembles the skin of organic matter on the bedrock of the planet, yet it is a highly detailed mineral adding to its ability to address complex skin problems. Adularia is strong and absorptive, so it is good to keep it cleansed. Also good in reducing fungi. It may help to regulate body-fluid balance, especially in hypertension and high blood pressure. *(4th chakra and all skin)*

AGATE: All forms of agate found worldwide possess the same chemical structure as clear quartz, silicon dioxide, however the growth structure of agate is called microcrystalline. Other types of quartz-based rocks and minerals are also microcrystalline, namely onyx, flint and jasper. All of these varieties are classed under the heading of chalcedony quartz. Agates contain other elements, such as iron, aluminum, or sulfur which cause their individual and distinct characteristics to make identifying their locality of origin quite easy. Because of this, some forms of agate are classified as semi-precious, such as Ellensburg Blue, because its original deposit was spread out by glaciers. Agates can be costly due to rarity — or they can be very affordable. This affordability makes agates an excellent option to balance the subtle, physical, mental and emotional bodies. Since agates are so distinct, they provide a wide variety of resonance frequencies for individual humans, animals, plants and localities. Agate is the state gem of Louisiana. Lake Superior agate is the state gem of Minnesota. Sapphire agate is the state gem of Montana. Blue agate is the state gem of Nebraska. Fairburn agate is the state gem of South Dakota.

Angelskin - What a perfect name for such a sublime mineral! Angelskin is easily one of the highest vibration agates. The subtle blending of lavender and rose provides an excellent foundation for higher ideals of love, patience and trust to blend charismatically with the heart, mind and body. When this occurs, it can be felt on the skin as a rush of goose bumps. *(4th and 6th chakras)*

Blue Lace - Use on the third eye area between eyebrows as a sedative for nerves and to relieve non-allergenic, tension headaches. It aids flow of expression and ideas by linking the throat and third eye chakras. Blue lace is an agate which addresses subtle body transmissions to the physical, and increases out-of-body awareness and experiences. Its highest character resonates with spirit guide and angelic connections rather than physical health issues. However, it influences the physical body effectively through body-mind healing practices of visualization, biofeedback and meditation. Any health issue can be alleviated and progressively released through its effects on the subtle bodies. *(5th toward 6th chakras)*

Botswana (Botswana, Africa) - This form of agate is found only in Botswana, Africa. Adds strength and emphasis to other stones in jewelry as it acts as an amplifier in harmony with associated colors. Aids oxygenation of body parts. The characteristic swirl provides a visual connection to the plant kingdom, wood in particular. All varieties are feminine power stones and excellent healers, and are recommended as personal power stones. Stones which are predominantly pink, white and opaque draw in celestial influences, such as angels and archangels, and the air element. These stones aid with physical issues which require dryness to heal and may appear to be impossible for human technology to solve. Dominantly bluish-grey, and more translucent stones draw in water element influences, such as evapotranspiration, clouds, fog, mist, waterfalls, gentle water bodies and water spirits. These stones aid with physical issues which require moisture for cleansing, and carry the mystical, magical healing properties of water. Stones with both colors are conducive for emotional, spiritual, and mental issues. Botswana agate supports the transmission of celestial or spiritual information through the action of water coming to earth or the physical. An excellent stone for arborists, foresters and those who work with trees, fountains, feng shui and devas. Botswana agate is listed in the Spiritual Apothecary in the Etheric Conditioner category. *(All chakras, varies with color)*

Bull's Eye (South Dakota) - This stone is from the Badlands of South Dakota, and is red and brown. Often given at the time of a first vision quest, but not commonly for vision quests all through life, this stone is believed to strengthen spiritual vision and to enhance receptivity to the guidance of spirit allies. It was occasionally given to newlyweds with the hope of supporting their role as parents of the community's next seer or visionary. The implication became a profound reminder to the parents that their community called for the blessing to be shared by all and not to be received as a personal accolade. Responsibility for raising such a child most often became a community effort. *(1st chakra)*

Burro Creek (Arizona) - This agate is found in light to dark lavender nodules, often dendritic. This quality of agate facilitates the completion of forgiveness, especially in matters of principle, justice, spirituality, philosophy and social order. Rather than being attuned to issues of intimacy as heliodor (lime green) is, Burro Creek agate with its lavender clarity resonates with the undifferentiated whole and bridges beings in a shared, overarching, group mind. Burro Creek agate and heliodor work well together. Burro Creek agate embodies the feminine aspect of beaver energy. See Unakite. *(7th and 12th chakras)*

Crazy Lace (Mexico) - This agate form originates only in Mexico. It is made up of quartz and hematite with lace-like curls of jasper. Crazy lace agate is a gentle absorber of emotional pain and should be cleansed frequently in sun, moon or rain when in use. It has so much variability that it is capable of aiding the variety of human frequencies which exist. *(All chakras, varies with color)*

Dendritic - Another visualization tool as one studies the intricate patters of mineralization. The black, coral-like (manganese) patterns are the stamp of its cousin kingdom. It is a past-life regression assistor. It connects the physical with the etheric so it is a marvelously accessible mineral for connecting with the higher self, unconscious and subconscious. Large pieces are excellent for rooms where hypnosis sessions, past life readings, soul retrievals and journeying occur. *(1st and 7th chakras, others depending on host agate)*

Dryhead - The pastel red and pink hues are trimmed with gray and white borders making these most beautiful nodules. Eases pain of sickle cell anemia. *(1st chakra)*

Ellensburg Blue (Washington) - This agate is pale to deep blue/lavender. Ellensburg blue was classified as a gemstone by Tiffany's many years ago before other forms of blue agate were discovered. Its rarity and potentially high quality blue with lavender undertones still sets it apart from most blue agates in the world. The origin of the stone is unknown as it is found in the end margin of glacial deposits from the last ice age in Kittitas Valley, central Washington. The opalescent quality tones the brow and throat chakras to relieve pollen headaches, sinus allergies, sinusitis, and migraines. Ellensburg blue epitomizes the essence of wild turkey medicine, the giveaway eagle. The valley where it is found is renowned as a summer gathering area for local tribes where potlatch ceremonies occurred. It is known as the valley of peace and harmony. Ellensburg blue embodies the energy of service and self-transcendence as shown by St. Francis of Asissi and Mother Teresa. Know that wearing it encourages voluntary simplicity, giving oneself and one's possessions to others selflessly, and humility in actions. It works optimally in association with programmed, clear quartz. *(5th toward 6th chakras)*

Evergreen (Washington) - This army-green agate formed in a narrow strip often with a milky white agate crystallizing to quartz on one side. It can look very much like fir needles embedded in snow. This is a relatively new material found east of Mount Rainier in Washington. It fosters simultaneously qualities of leadership and teamwork, thereby enhancing the attainment of group will and mind with all members equally sharing responsibilities of management and supervision. It facilitates ease of transition of leadership between group members, cooperation, and recognition of skills. *(4th through 7th)*

Fire - Balls of brown iron areas are deposited in milky chalcedony nodules. Fire agate stimulates the whole body system. Good memory stone and tissue regenerator. It balances, grounds, and protects those doing "light" work. Earth and fire combined into mineral. Emphasizes fire areas of a person's astrological makeup. Fire agate is an Etheric Cleanser in the Spiritual Apothecary. *(1st and 6th chakra)*

Glory Blue (Montana) - This stone regulates the flow of white light energy through the neck and throat chakra. As light enters the subtle bodies and flows through the physical body, the muscles and bones of the neck can become rigid in response to the immense pressure experienced there. This area of the body carries the greatest stress when moving light energy, and glory blue agate assists in relaxing this area for transmission. A rich blue, somewhat purple color. *(6th chakra)*

Holly Blue (Oregon) - The old timers' name for holly blue is Callipoolia blue for the river it was originally found near in Oregon. This stone has several qualities which are not necessarily parallel or complementary. They are at angles to one another, creating sacredness through interdimensional connectedness. The stone is excellent for fish aquaria and fountains, or for offering up as a symbol of connectedness to the Piscean Age now past. It is an exceptional altar stone as it defines sacred space, and can be placed on the dining table to represent thankfulness for sustenance received from the One. It is a purifier of air and water, and can be cleansed in rose water once a week for several hours. There are some ineffable aspects to this stone which will come with use, and it is important that this stone have an experiential base in the individual so that the sacredness which is taught through and embodied in this stone is made real and palpable. Holly blue is listed as an Auric Conditioner in the Spiritual Apothecary. See Chalcedony (blue/purple). *(6th chakra)*

Iris - Thin slices reveal the characteristic rainbows when tilted in the light. Microscopic studies have revealed there can be up to 10,000 layers of agate per inch in this tightly banded stone. Regenerates loss of touch in fingers that have suffered burns or injuries (30-minute applications). Promotes receptivity to inner truths. Its rainbow colors can activate other chakras. *(Hands, rainbow colors activate other chakras)*

Laguna (Mexico) - Laguna agate is usually a creamy red/pink and gray colored fortification agate where the banding is angled to points and scallops rather than rounded. It bears stunning colors and a complex pattern which usually commands a high price for an agate. Laguna agate can be called the storyteller's stone, because it was used to highlight a segment of story. The story teller passed pieces of stone, usually forms of agate picked up during travels and trades, around the group of listeners to imprint visual images which helped them recall the story line. Laguna agate was prized as a talisman of "the fire within," a time of vision questing, or realization of life purpose being described for one of the characters in a story. It was not unusual for a storyteller to give this stone to a young person as an affirmation of storytelling abilities. *(4th chakra)*

Moss - Moss agate improves circulation in the blood and lymphatic systems. Digestion and elimination are improved. Harmony is a key attribute. The green inclusions form a link to nature for our subconscious. It is a natural stimulator of the meridians of the left thumb (white) and little finger (green). It is a crown and heart chakra connector which facilitates heart-embedded meditation and daily manifestation of compassion in action. *(4th and 7th chakras)*

Picture (Montana) - A popular variety of picture agate is found in Montana, USA. It usually bears black dendrites in milky to clearish agate with some other colors occasionally mixed in. It is prized as an agate which readily offers landscape pictures and some animal shapes. Meditation, focus or concentration on this stone improves coordination of left and right brain functions. It can aid creative visualization. Thin slices are excellent for color healing applications by allowing natural or full spectrum light to shine through them. *(Lower chakras)*

Pine Moss - The gentle presence of pine moss agate is excellent for calming disturbances in the nervous system. It is especially attractive to children. This red/brown moss, when combined with green moss agate, provides a powerful heart chakra connection. *(1st chakra)*

Plasma (San Benito County, California) - This agate contains nickel, mercury and cinnabar. Cinnabar is excellent for conditioning reflexes, and mercury (or quicksilver) enhances reflex and muscle response. Nickel enhances frequency perception and clairaudience. As a combination of minerals, this agate supports shapeshifting, moving core energy out of the second chakra to other points outside the body, and using the body in meditative movement. *(1st, 4th and 7th chakras)*

Plume - Besides the possible visual images which may be conjured from the foliage-like weave pattern of mineralization, the brown, red, and orange color range works on the regenerative and other lower vital organs. Pants-pocket carrying or similar placements can energize and activate healing flows in navel and base chakra areas. Grounded, expressive power would be the complete theme of these stones. *(1st and 2nd chakras)*

Polka Dot - A cellular-level healing stone. Strengthens osmotic properties of cell walls and strengthens cell membranes. Compatible with Vitamin C activities in the body. Strengthens the immune system. It occurs white with yellow/brown spots or blue/gray with spots. Indians used to heat rock walls with fire then splash water on the jasper to split off chips for arrowheads, eventually creating a cave. *(1st, 3rd and 7th chakras)*

Polyhedron - These unusual angular-shaped geodes look like stone origami, and are prized for their polyhedron uniformity. They occur only in Brazil. Polyhedron agate relaxes and soothes tissues and cells around and in the spinal column. It encourages reduction of swelling. It promotes relaxation, and is an excellent choice to relieve ergonomic stress and tension from work at a desk or communications terminal. It is good for the shoulders and feet, providing energetic effect similar to reflexology. This agate is a good meridian alignment mineral. *(Etheric chakras)*

Rattlesnakeskin - This is an unusual and rare form of snakeskin agate which forms as white-skinned, milky agate nodules. It can be found with a leather-like skin pocket on one end of the agate with a free stone inside. It rattles when shaken and is good for ceremonial use.

Snakeskin - Promotes inner peace. Wearer spreads joy to others. Apply lighter version to palms and soles of feet, whereas the dark snakeskin is applied to brow and temples (15-minute applications). *(Hands and feet)*

Strawberry - This agate usually forms massive and opaque with a crisp, rich pink color. Strawberry agate is excellent to use for major life transitions such as birth and death as it helps lead the soul across physical "barriers." It also helps in making irrevocable life decisions. It facilitates higher self transmissions and spirit guide interaction. *(1st to 4th chakras)*

Turritella - Turritella agate is a fossil shell replacement. It is grounding and supports individuals lacking adequate earth signs in their zodiac. It is a water/earth vibration and synthesizes these aspects beautifully. It is a protection mineral for people engaged in water oriented activities or jobs. *(1st chakra)*

Zebra - White with black, jagged strips, it's well suited to the name. A mineral of neutrality. Zebra agate is excellent for individuals with overly aggressive natures, good for temperamental or violent people needing balance and allowance in their lives. A good stone to place near the sleeping area for these individuals or anyone desiring a good night's rest. An excellent application is in prisons, reform schools, etc. *(1st and 7th chakras)*

AJOITE (Copper Phantom in Quartz): Ajoite is the "Phoenix Rising" stone. Ajoite is a hydrated copper aluminum silicate with potassium and sodium. The quartz points bear bright, turquoise-included phantoms and occasional red iron inclusions. It is a master integrator specifically addressing issues of discord such as bigotry, prejudice, land disputes and religious differences. It is no coincidence that ajoite-included quartz was mined in South Africa, as the higher purpose of ajoite is to bring light to the resolution of racial issues and genocide.

The image of the Phoenix rising refers to the spiritual evolution of humanity as a whole. The Oriental and African cultures achieved spiritual obstacles in their earth lessons through Lemuria (Asia) and Atlantis (Africa). It has been the work of European cultures for most of this century and most of the last to integrate diverse spiritual paths with changes in perception triggered by quantum physics and the nuclear bomb. Ajoite provides the necessary mineral complement for this transformation work to proceed peacefully and successfully. Ajoite sheds light on issues of shame and shame-based beliefs. Racial bigotry, in particular, is a shame-based problem which ajoite stimulates to resolution.

Zecharia Sitchin's series of books on the origins of the human species, beginning with *The Twelfth Planet*, is a profound resource for understanding Sitchin's theory that genetic engineering of *Homo erectus* and non-terrestrial beings led to the development of proto-man in South Africa. The first humans were engineered to work in the mines of South Africa, and were deliberately engineered without the ability to sexually reproduce for convenience and control. When future human versions were able to reproduce, humans were thrown out of "the garden," and forced to survive alone. They were spread around the planet to reduce their ability to form alliances. This experience grounded shame deep into the genetic fiber of humans kindling all manner of social strife and ill. The very locale where humans were "born," enslaved and thrown to the wind is also the site of a high mineral ally. Earth does come to the rescue!

The time for rising in the flames and smoke of a dying civilization is upon Earth, and ajoite has made its appearance none too soon. In the last century, four spiritual leaders have expressed the intention which ajoite crystallized in its matrix. The first was a Nez Perce Native American by the name of Chief Joseph. The second was the writer and social activist Henry David Thoreau. The third was Mohandas Gandhi who practiced law in South Africa for a time, and the fourth was African American Martin Luther King, Jr. Their message was nonviolent resistance. "I will fight no more forever," as Chief Joseph said.

In every case, the oppressor these men resisted was of European origin. Europeans have incorporated the "sin" aspect of being rejected by our genetic engineers more than other cultures. This century has put this technologically and scientifically oriented culture to the test. We chose to use the atom bomb to stop World War II, which triggered a holocaust of world changes which threatens every living thing on the planet. The ultimate recognition of the nonviolent ideal is yet to be expressed by Oriental cultures; however, the Japanese have been viscerally altered by their experiences in Hiroshima and Nagasaki. Their embracing of this ideal will signal the ascent of the Phoenix. See Papagoite-Included Quartz and Ocean Picture stone. *(6th through 12th chakras)*

ALABASTER: Alabaster is a compact form of gypsum, and is thought to derive from alteration of anhydrite. All types are conducive to carving as the material is not excessively hard. More absorptive energetically than marble, alabaster carvings are excellent for ceremonial areas where they receive regular exposures to incense and smudge. Do not soak alabaster for cleansing. The previously mentioned forms of cleansing are adequate as is sound cleansing with bells, gongs, chanting and singing. Alabaster is a Healer in the Spiritual Apothecary.

Orange/Peach - This form of alabaster is especially soothing to the gastrointestinal tract and genital organs, aiding conditions, infections or diseases of that area, such as ulcers, diverticulitis, colitis, cancer, Crohn's disease, endometriosis, candida, cysts, and fibroids. This tone of alabaster aids in oxygenation of the blood and removal of free radicals from the bloodstream. This form of alabaster brings in upper chakra energies for a more finely balanced whole body state and awareness. *(1st, 2nd and 3rd chakras)*

White - A milky white stone revered in Biblical times for its purity and representation of high ideals. This mineral facilitates sensitivity and inner assurance when used in meditation, healing, and psychic readings. Its softness also allows peaceful thoughts to be carried outside of the self into action and for auric healing to be absorbed into the physical body. *(7th chakra)*

ALBITE: Albite is a sodium aluminosilicate classified in the feldspar group. A complete series from albite to anorthite forms the plagioclase series in this group. It is related to labradorite, microcline, orthoclase and sanidine. It is white or clear and refractive. Albite is often found associated with other minerals, and offers a free-flowing element to the mix which increases impact and effectiveness. It is important for ceramics. This is a good mineral to counteract heat stroke and cramps. It aids all fluid issues in the body from saliva to intestines. It is valuable to maintain cellular hydration in all strenuous activities impacting the muscles and nervous system, such as sports and intense meditation or psychic work. *(All chakras)*

ALEXANDRITE (Gem Chrysoberyl): Alexandrite, or gem quality chrysoberyl, is beryllium aluminum oxide. Low grade specimens of alexandrite can be black or green. As clarity increases, its green nature changes to blue, red or purple depending on the light source. This mineral has overall healing impact. Healing on all levels is instigated by alexandrite, because it is a serpent mineral due to the presence of beryllium. It works on the nervous system and some of the key glands, such as the spleen and pancreas to supply energy for psychic work. It is unsurpassed as a gemstone which stimulates prophetic visions especially of a transformative nature.

With intention, the highest octave pulsing from alexandrite is eagle energy, spirit eyesight, the dual vision of the chosen and the damned. It resonates with the planet Pluto and the astrological sign of Scorpio, and wearing it is a choice of all choices. This resonance makes it a stone which can be used consciously to decrease earthquake seismicity, and it alleviates nervous system stress from prior awareness of an impending earthquake. For some people, alexandrite is all there is, all that needs to be worn or worked with. So many stones, so little time is not a worry for people who resonate to alexandrite. Some cultures consider it a stone of luck and prosperity. It is most commonly lab grown now as it is quite rare in nature. Alexandrite is the June birthstone. *(2nd, 7th and 8th chakras)*

ALLANITE: Allanite is a hydrous silicate categorized in the epidote group along with zoisite. It is a white clay with a brick red, zig-zag pattern forming in seams. Enhances the recognition and storage of fringe perception, those sensory stimuli which are screened by the mind to prevent sensory overload. When used appropriately, it will provide data for day dreaming or sleep dreaming which can be beneficial to all, especially creative types. *(1st and 7th chakras)*

AMAZONITE (Microcline): Amazonite is a potassium aluminum silicate. It is a pale to rich turquoise-colored form of microcline that is visually soothing. It synchronizes the heart, throat and third eye chakras. It activates thought energy for completely balanced expression and manifestation of ideas and impulses. Amazonite counteracts psychic burnout from the pressure of white light energy impinging on the narrow throat chakra and flowing from the crown to the trunk or the flow of earth or kundalini energy up from the base chakra and out of the body. Acts as a facilitator and lubricant energetically, so it is excellent for the nervous system. It can also act to relieve bronchial spasm. Tending toward its greener expression, amazonite is a stone of joy. *(5th chakra)*

AMBER: Amber is a mixture of plant hydrocarbons, specifically fossilized or petrified tree pitch. Depending on its place of origin, amber will be softer or harder, because it is more or less petrified. Famous localities include the Baltic region (90 percent of the world's supply) and the Dominican Republic. The English word electricity derives from the Greek word for amber, *elektron*, because when amber is rubbed, it builds up static charge which attracts other objects. Amber is most typically yellow to dark orange, but it can be white, creamy, or red, and fluorescence causes green and blue shades. Amber is so light that it will sink in a supersaturated solution of salt which helps to distinguish it from imitations which can be made of plastic, glass and manmade resin.

Greek mythology says that amber derives from the tears of the sisters of Phaeton, son of the sun god Helios, who was killed by lightning. It is also said that Apollo wept tears of amber when he was driven from Olympus. Amber has been highly regarded as a rosary stone for centuries, and at its height of interest as a carvable stone in the 16th and 17th centuries, it was made into chalices, chandeliers, and religious sculpture. Amber is often included with insects and other organic matter.

Amber strengthens tissues of the brain and nervous system, and draws toxins from body tissues. It purifies the etheric body by absorbing negative energy. Cleanse regularly with smudge, incense and air baths. Don't cleanse amber in salt water. Baltic amber is particularly rich in holistic healing ability. With deep orange hues and black inclusions, it is even grounding. Amber is always excellent for centering as it purifies the subtle bodies much as smudge does, but it is absorptive which requires cleansing attention on a periodic basis. In a ritual of release and cleansing, amber absorbs meditative thought and physical transmissions. It can be used to absorb trauma issues which need to be transmuted to the Light. Burn the piece in an outdoor location. Younger amber has been a form of incense for many centuries. Amber is an Etheric Cleanser in the Spiritual Apothecary. *(2nd and 3rd chakras)*

AMBLYGONITE: Amblygonite is a hydrous lithium aluminum phosphate, colored from white to light blue, and found in Brazil and South Dakota. It can be both massive and grainy or crystallized which can look slightly yellow. It is a soothing and healing mineral which combines the qualities of lithium and phosphorus to help the nervous system. For people who are very active, nervous and have trouble keeping weight on, amblygonite provides the necessary elements to calm nerves and oversensitivity. It is a good companion stone to porphyry, and is a useful stone to have in offices of intense mental activity, such as law, architecture, engineering, research or libraries, as well as places which are full stress and deadlines. It is an important mineral for the bones, and when complemented by a mineral containing calcium, amblygonite's attributes are better utilized.

On its own, amblygonite works well with mild anxiety symptoms. Combine with richterite for increased efficacy, (see Richterite). They work well together to prevent seizures. Experiment with amblygonite alone to assist with earthquake sensitivity, and add other minerals to assist with nervous system issues as needed. It is a stone which impels the prophetic voice through ecstatic recognition, and may be called "the Prophet stone." *(5th to 7th chakras)*

AMETHYST: Amethyst quartz results from exposure to natural or artificial radiation, which causes minute iron and aluminum impurities to expose themselves. Impurities need not be more than one in one thousand silicon atoms for the color to be evident. Amethyst is the most expensive and highly regarded color in the quartz family. Amethyst gained its color, according to Greek myth, when the god Bacchus promised, after being insulted, that he would consume the first person he saw. The hapless person turned out to be Amethyst on her way to worship the Goddess Diana, who swept down to save her and turned her into clear quartz crystal. Seeing the error of his anger, Bacchus poured grape juice over her, turning her stone body purple. Based on that legend, people believed amethyst to be a cure for drunkenness, anger, disturbed passions and frustration. It is the stone of temperance and sobriety, and is said to stimulate female sexual glands.

Amethyst assists body/mind surrender to and integration of higher spiritual energies. It stills, opens and focuses the mind during meditation, mediation, counseling, and intellectual creativity. It provides calm for sleep with inspired dreams, and can relax sleeplessness due to anxiety, stress, burnout, feelings of being overwhelmed and depression.

Amethyst allows the intellect and ego to surrender to the amethystine ray. The Romans used amethyst in wine to counteract drunkeness. It is effective in alleviating and shifting compulsive disorders, mental illness, addictions, negative thought patterns, and physical illnesses due to stress and the previously mentioned areas. It counteracts excessive behaviors. It reduces bitterness, resentment and harsh tones in speech by aligning individuals with their highest good. Over a period of years, it completely shifts patterns of negativity learned from childhood, allowing the reemergence of the inner child and the healthy inner parents. Therefore, it resonates to triune archetypes and symbolism of the father, mother, child and the tetrahedron which is the primary order geometric expression of life. It encourages feelings and expression of tolerance, acceptance and detachment as well as gradual awareness of sufficiency. This means that amethyst is so effective at connecting physical to spiritual that one eventually regards what one has as enough. Decreasing needs for over-consumption lead to feelings of basking in love, light and one consciousness.

Amethyst is a stone which encourages attainment of understanding the unified field theory, the union of souls and the oneness of all universal energy. Christians often view amethyst as a stone of humility, because of its hue. Amethyst activates and strengthens the immune system (through feelings of calm, peace, safety and serenity in the Light), the endocrine system, and the pineal and pituitary glands. Its calming vibration slows heartrate to decrease movement of poison in the body. It deepens intuition, higher self connection, and past life release of karma. It cleanses all the bodies and the physical and etheric environments through absorption of negativity, but it also charges all the bodies with high healing energy. It is an essential stone for all forms of healing work. Amethyst is the State gem of South Carolina. Amethyst is the February birthstone. See Aurichalcite. All forms of amethyst are categorized as Auric Conditioners in the Spiritual Apothecary. *(6th and 7th chakras)*

Though all amethyst has a common impact, there are some subtle differences in the following varieties:

African Chevron - Typically found massive, this is a very dark amethyst with a white chevron phantom which is good for cutting cabochons for jewelry. It is excellent as a massive stone for placing in healing rooms and on the body during healing, because it brings in both the white and amethystine frequencies of light. It is good for remote viewing applications as it promotes accountability on the highest frequency of spiritual integrity, thereby counteracting the temptation to abuse remote viewing ability for power, financial gain, revenge or control. It is such an activator on a high consciousness level that it is not conducive to everyday mundane activities or staying present for work or driving for those people who are already highly open and intuitive. For those people who are denser in their lifestyle and physical awareness, chevron amethyst can stimulate noticeable rapid shifts toward higher energies. *(6th and 7th chakras)*

Brazil - From light to medium dark color, but always a stubby proportion. Used for faceting, etc. Best in clusters for storing and slowly releasing energy to an environment. *(6th chakra)*

Central Mexico Phantom (Guanajuato, Mexico) - A subtler "shunted" effect. Like all phantoms, they release energy in packets. *(6th and 7th chakras)*

Chlorite-Included (Africa) - "Earthlight stone." This stone is the harbinger of Pleiadean dreamtime as it melds the purple and green rays in quartz, and harkens to the wearing of purple and green in the Pleiades, which denoted an individual as a healer and a peacemaker. It is the first choice for the generator crystal in cosmotron or tonelight technology. Stonecyphers prefer this mineral combination over all other purple and green combinations for deep trance and meditative work. This aspect of amethyst addresses nerve-oriented diseases such as Parkinson's and neuralgia. *(4th, 6th and 7th through 12th)*

Siberian - Though actually mostly dark purple, it has phantom areas, usually blue, alternating with the amethyst color. *(6th chakra)*

Thunder Bay (Ontario, Canada) - This classic locality produces beautiful amethyst with a red hematite inclusion that is easily identified. It connects the first and seventh chakras. It is a polarity balancer, bridging aspects of peace of mind with grounded alertness. On this continent, it is unequaled as a meditation stone which calms the mind and deepens the still mind connection to the root chakra. It enhances recall and integration of meditative experiences. *(1st and 6th chakras)*

United States - A number of localities have produced amethyst over the years. Many are lighter in color. Local vibrations imbue varieties with specific attunements. *(6th chakra)*

Uruguay - This is typically the darkest and most expensive cluster-type amethyst available, and is highly prized for metaphysical use and by collectors. It usually has a thinner matrix than Brazilian amethyst. It is rarely found in geodes. *(6th chakra)*

Vera Cruz (Mexico) - This form of amethyst is often pristine in clarity and colored a light lavender at the tips fading to clear at the growth base. It is often found double-terminated, making it excellent for healing applications and jewelry. It is highly regarded for natural point jewelry. Besides the esthetics of good color and sheen, the longer, thinner points transmit energy well. *(6th chakra)*

ANALCIME: Analcime is a hydrated sodium aluminum silicate in the zeolite family of minerals which looks like milky-white garnet crystals. Analcime stimulates an ability in humans called synchronous drift. It allows individuals to maintain an agreed upon distance or intimacy when they will be separated for lengthy periods of time and exposed to different influences during that time. If individuals do long distance healing, meditation or dreaming, it allows brain wave patterns to move in synchrony. Analcime is a fourth dimension mineral with many high qualities for people seeking to expand their potential. *(7th chakra)*

ANAPAITE: Anapaite is a hydrated calcium iron phosphate often found as a botryoidal coating in a clam fossil from Russia. It is a fresh green mineral tending toward clarity. It is an overall tonic for the body, auric and etheric. It raises healthy tissue to super-energized levels which can support regeneration of areas debilitated because of injury or cancer, etc. *(All chakras)*

ANASAZI RED OCHER (Sedona, AZ): Anasazi red ocher is from a place which is considered the womb of Earth, Sedona, Arizona. When dry, it is a messy red, powdery clay. It resembles the red monolith, Ayers Rock (Uluru), which is the heart of Earth. Red ocher leaves a stain like menses, and is very sacred. Its highest purpose is to be used in rituals which honor Earth, and when taken to ravaged lands and deposited back to her, it provides lifeblood and nourishment, It is a potent antigen for the Earth, allowing her to replicate bacterial antibodies to combat pollution, and to neutralize positive ions and negative energy fields. Red ocher is a powerful ritual healing tool for the reproductive organs. It embodies the power of aggressive healing, therefore it resonates to badger medicine. *(1st chakra)*

ANATASE: Anatase is a titanium dioxide, related to rutile, and is excellent as a lubricant and fluidity enhancer to promote smooth operation of machines. It aids humans in understanding complex mechanics, and helps machines to be comprehensible. *(1st chakra)*

ANDALUSITE (Gemstone Variety): Andalusite is a simple aluminum silicate with an olive green to brown color tending to excellent clarity in gemstone quality. It is trimorphic with kyanite and sillimanite. These three minerals belong to a category which can be called, "I'll believe it when I see it." All three in their own manner facilitate physical visual and aural receptivity to subtle levels of invisible frequencies, thereby increasing the ability to observe auric energies, past, future or distant events, internal states of living organisms, and sound emanations. Andalusite is a wonderful stone for people with either or both fire and earth in their astrology charts. It is an earth healer stone meaning that it supports absorption of healing energies from the earth as well as efforts to send energies to earth. It is an earth speaker stone and an excellent spokesman for what Earth makes clear that she needs to be healthy and balanced. Andalusite supports geobiological sensitivity. It also increases the ability to maximize vision globally and manifest locally. See Chiastolite (Cross Stone). *(1st through 4th chakras)*

ANGELITE: "Herstone, The Goddess Stone." This mineral is found in Peru as a massive, light blue nodule with white skin. When the Mother speaks, her children listen. All earth events such as volcanic eruptions, tornadoes, wind storms, droughts or floods are expressions of the Mother's being. Yet she is also a crooner, singer and whisperer of old stories and truths. We are aware of her tremendous voice, but it is her voice at the edge of silence and in the heart of life that we miss, have difficulty hearing, wish to honor, but doubt our ability to hear. The Great Mother always knows when to speak. From the spirit side first comes knowing and awareness, then seeing and hearing, then speaking. Emotion results from all this sensation and integration with the higher self and finally action and creation.

For humans, the opposite progression is true. We are created and forced through birth into action. Having just arrived from immersion in spirit, we are fully aware of our life purpose and we have much loving intention. We speak our needs and wishes before we fully integrate the meaning of sound and sight in the physical. Finally we become aware and capable of knowing as physical beings. Angelite allows humans to tune out our overwhelming physical imperatives in order to hear clearly the voice of spirit as spoken through the Great Mother. The throat, much more than the heart, is a serious integration point for the coming years of transformation, because it is at this point in the body where communication is most easily disrupted and perverted. The narrowest gorge in the river experiences the greatest force of flow as light moves through us from the crown chakra. Angelite strengthens the perception of whispers and truths even in the torrent of the river. *(5th chakra)*

ANGLESITE: Anglesite is a lead sulfate categorized mineralogically in the barite group. It occurs as tabular crystals with striations which come to a dull, arrowhead-shaped point, and with a color range from clear to yellow to golden. It was a highly prized mineral in trade by Africans and Mayans in past centuries (an indication that cross-Atlantic travel did occur and there was communication between the two continents prior to the 1400s). Anglesite, like golden barite and celestite, was considered a stone of truth. Anglesite is a stone of truth in action and at death. People who received this stone as a gift from royalty were considered to be most highly blessed. Anglesite was buried with the owner or given to persons of deep spiritual conviction prior to death or at the time of letting physical attachments go in order to make a spiritual commitment of poverty and chastity. *(3rd chakra and higher chakras)*

ANHYDRITE: Anhydrite is a simple calcium sulfate which can result metamorphically from the dehydration of gypsum. Alabaster is thought to result from alteration of anhydrite. Aesthetic baby blue "fronds" look more like plants than minerals. The finest specimens come from Naica, Mexico. Metaphysically, it is a Class "A" mineral. It acts something like a radar detector for areas which are storing toxins or negativities. The locales identified by anhydrite will either tingle or cause mild discomfort which can be massaged, worked energetically or harmonically, or meditated on to cause release. Anhydrite balances hormones which are upset by pollutants, toxins or stress. Anhydrite will locate unwanted energy leaks, entities and attachments. It is an "angel duster," meaning angels resonate very clearly with it, and are attracted to its use and presence. It is vibrationally in tune with dioptase and gypsum (selenite). Having both in the environment will create a relaxing sense of balance and attunement. Excellent for elixirs to address nervous system problems. For the nervous system, the light blue variety is superior to the white glossy specimens. White forms of anhydrite are better for energy work and entity removal. Anhydrite is placed in the Healers division of the Spiritual Apothecary. *(5th chakra and upper chakras)*

ANNABERGITE: Annabergite is a hydrated nickel arsenate, and is a member of the vivianite group along with erythrite and kottigite. Annabergite is a breathing mineral, excellent for pranayama, rebirthing, childbirth, meditation or breath control exercises for singing or oratory. It stimulates the chest region to open up to deeper breathing, and the solar plexus which has an effect on all systems in the body. Annabergite is a mineral which supports interconnected parts such as the lymphatic, nervous, and endocrine systems. It facilitates the understanding of complex interrelationships. *(3rd and 4th chakras)*

APACHE GOLD (Oriental Gold): Apache gold is a massive stone where the black is chert and the golden is a chalcopyrite blend. It comes from Arizona. It is an excellent mineral for artisans or any artist who creates three dimensional objects such as sculpture, pottery or jewelry, etc. It is highly oriented to kinetic energy, meaning it stimulates movement and motor functions in the body, especially between the brain and hands. It is applicable to all forms of healing work which involve the hands, such as massage therapy. It enhances spirit communication through and recall of dreams. It works on the aura to generate and increase sensitivity to surroundings or other people, so it is excellent for dowsers, paranormal researchers, psychic detectives, tour and museum guides, and workers in resorts, health spas, and on cruise ships. *(1st and upper chakras)*

APATITE: Apatite is the common name for an apatite group mineral which is a calcium phosphate with other anions present. The name of the type of apatite has a prefix, such as *fluor*, to indicate the predominant anion which in this case would be fluorine. Chlorapatite and hydroxylapatite indicate the presence of chlorine and hydroxyl anions, respectively. Fluorapatite is the most common variety. The skeletal, muscular, circulatory and nervous systems all benefit from apatite. Apatite is a primary constituent of bones and teeth. It joins calcite minerals in aiding the metabolism of food, therefore it increases the sense of satisfaction from food at the sufficiency level rather than the overconsumption level. The chemistry of apatite (calcium phosphate) clearly indicates a natural life-force bonding which is also found in several types of food, namely sesame seeds, which is vital for the immune and nervous systems. This mineral is recommended for people on the go with busy, tight schedules who are burning out their nervous systems. It is also excellent for those experiencing a loss of mental and physical strength or efficiency. All forms of apatite are found in the Auric Openers category of the Spiritual Apothecary.

Blue - Occurs in small, pencil-shaped crystals. This color opens the throat chakra, and reduces tension and stuttering. It improves verbal expression. *(5th chakra)*

Blue (deep) - This form of apatite occurs as thick, stubby pencil-shaped crystals and in larger, massive pockets which can be polished. Blue apatite, in ages past, was prized as a vision seeking mineral, and was considered an essential component of the Spiritual Apothecary. Trained priestesses developed this apothecary over a lifetime for use on common as well as adept folk. It was passed on through bloodline to the next priestess. Most apothecaries were destroyed over the centuries. Working with blue apatite can facilitate recognizing the necessary minerals for a personal spiritual apothecary, and it is a Healer in the Spiritual Apothecary. *(6th chakra)*

Blue on Clevelandite with Lepidolite - These apatites occur as small, light blue toothpick-sized crystals on fist-sized chunks of lepidolite. This powerful combination shows the inherent genius of mineralization on Earth, because this combination draws together three important areas of personal power: life purpose and the chain of past life history, vision questing and the development of a personal Spiritual Apothecary, and right use of will as the mineral Clevelandite connects humans to the Divine Will. There is no other more powerful combination for facilitating the quest to know self and to direct self upon the journey to the highest self-expression. It's an important mineral for counteracting osteomyelitis. *(5th through 7th chakras)*

Blue/Green (Russia) - This form of apatite is excellent for drawing energy to healing bones in the shoulders and upper chest. Its is a mineral said to be in support of the "watch tower," meaning it is especially related to people in occupations involved in watching and facilitating the movement of others, such as air traffic controllers, radar operators, forest rangers, etc. It is a powerful ally for photographers, movement therapists, dance teachers, and conductors. *(4th, 5th and 6th chakras)*

Golden - An excellent mineral for counteracting arthritis. It is more effective when used in combination with a green mineral such as dioptase or any mineral which alleviates depression. Golden apatite is specifically useful for any bone-related problem or illness. Good for past life recall. *(3rd chakra)*

Green - As with all the apatites, green apatite is good to support bone mending, restoration and healing. The green ray is especially tuned to the bones between the navel and neck. For people with past lives as healing practitioners, green apatite stimulates recall and empowerment of healing skills and practices. *(4th chakra)*

Purple - Beautiful, lustrous, stubby crystals have been found in Panasquera, Portugal. This type of apatite is excellent for healing bone tissue from shoulders and above. It also aids recall and empowerment of psychic skills and practices from past lives. Energizes the third eye and frontal lobe of the brain. *(5th and 6th chakras)*

Yellow - The classic location for these short, yellow pencil-shaped crystals has been Durango, Mexico. This form of apatite strengthens and regenerates muscle and bone tissue by improved calcium and vitamin C absorption. Helps with willpower.

Its compact power, like a battery, activates other minerals and ourselves. This form of apatite aids connection to past lives. Yellow apatite is found in the Auric Openers category of the Spiritual Apothecary. *(3rd chakra)*

APOPHYLLITE: Apophyllite is the common name for fluorapophyllite, which is hydrated potassium calcium silicate. It is often found associated with the zeolite mineral, stilbite. It is either clear, white or light green, and forms in square, striated crystals with a pyramidal termination which can be pointed or flatly truncated. The finest examples come from Poona, India.

Apophyllite is a healer's stone, first and foremost. It is categorized with other important minerals in the Auric Cleanser division of the Spiritual Apothecary. It principally effects and operates on the auric field of animal bodies. Therefore, its physical effects of supporting body and tissue fluid balance, the liver and the kidneys is produced through resonance and transmission from the auric field. It is a stone which immediately alters conscious awareness into intuitive sensing, but requires time for physical effects to be noticed. It is effective in all activities where heightened intuitive awareness is required, and is not recommended for daily wearing.

As a dimorphic, perceptual analysis tool, apophyllite connects this realm of reality with extraterrestrial realities. The term, dimorphic perceptual analyzer, is used in the technical jargon of extraterrestrial space pilots, such as ones from Arcturus, to refer specifically to the shift in consciousness which occurs between dimensions during space travel. There is a haze filter effect to this mineral, which is excellent to counteract communications static as well as visual and intuitive fogginess caused by fatigue. This is the dimorphic quality — its "two-shape" essence. To enhance concentration, apophyllite, or the dimorphic perceptual analyzer, allows unnecessary information to be taken out of a given space continuum which is the same as filtration or static reduction. This mineral is highly prized by space travelers of the truly interdimensional genre rather than the imaginary/conceptual/spiritual genres. *(7th and upper chakras)*

AQUAMARINE (Transparent Blue Beryl): Aquamarine is a blue beryllium aluminum silicate. Aquamarine balances the functions of the thyroid and pituitary glands. It helps to transmute accumulated toxins in the sinuses, throat and upper chest. Among the beryl family of minerals, aquamarine is found in the Spiritual Apothecary as an Auric Conditioner. Once the subtle and physical bodies have been cleansed, opened and energetically healed, aquamarine stimulates retention of healing properties. Historically prized by mariners and sea voyagers as a protection stone, aquamarine supports staying calm and aware during times of danger and crisis. It resonates to water energy. It transmits soothing energy to the survival center of the brain where it most ably supports people to replenish composure after crisis and difficult times. It enhances unflappable centeredness and presence through bridging white light frequencies to the physical body. During times of spiritual seeking, aquamarine stimulates higher self connection to the inner self, and brings the inner self into harmony with universal truths. Then it supports communication of these universal energies through physical movement such as dance, art, laying on of hands, singing, chanting, toning, and inspired witness. Aquamarine embodies the energy of aquatic or water dragon. Aquamarine is the state gem of Colorado. Aquamarine is the March birthstone. *(5th chakra)*

ARAGONITE: Related in geologic processes to alabaster and travertine, aragonite is a simple calcium carbonate classified in the aragonite group along with cerussite (lead carbonate) and witherite (barium carbonate). It is chemically related or dimorphous to calcite (which is also calcium carbonate), and is a higher order energetic expression of calcite properties in its particular color of expression. Aragonite crystallizes in orthorhombic patterns while calcite crystallizes in hexagonal. Aragonite is found in the Etheric Sealers category of the Spiritual Apothecary.

Blue (Laurium from Greece) - This massive form of aragonite possesses a stunning ice-blue color which applies directly to the lymph nodes of the throat chakra aiding the immune system. It can be included with dendritic layers. It is a stone of appraisal, evaluation and governance which supports the roles of leadership, rulership and judgement. It modulates the movement of white, lavender and cobalt light energies into the lower chakras through the throat. It is a stone of empowerment through listening rather than speaking, so it enhances receptivity to others and understanding of their needs. While the golden form of aragonite supports lower chakra receptivity to light, this form of light blue aragonite infuses the lower mind, the survival center of the brain, with universal life force energy. This can lead to euphoric states of awe, wonder and joy which will sedate pain and depression. *(5th chakra)*

Brown - Morocco produces drum-shaped single crystals which cluster together in bundles. It looks like a pollen molecule or a dandelion head gone-to-seed. It is normally a clear aragonite with brick brown inclusions, therefore it links the crown, first and second chakras. This combination is a good tool for biofeedback, visualizing health recovery, behavioral modification, and body/mind integration. It helps body memory of complex disciplines of body movement from dance to martial arts to tai chi. It synchronizes personal ability with the needs of the community, divesting an aura of social grace and good manners. *(1st, 2nd and 7th chakras)*

Copper Sulfate Coating on Aragonite - This specimen is clear or white with a dusting of rich, blue copper sulfate. The properties afforded by this combination bridge masculine (universal) and feminine (earth) life forces with the basic instinct to form higher organisms, from mollusks to humans. It's literally like the upward and downward drafts of kundalini flow sizzling along the spinal column, and it is a signature mineral for all metaphors and mythologies regarding the origin of sentient beings, especially humans. While it resonates to the nervous system, blood, and bones, it is so electrically stimulating that it is not at all grounding. *(5th and 7th chakras and spinal column)*

Cuprian - Cuprian aragonite forms as a sky blue crust with a fibrous, refractive edge. Copper ions are deposited in the matrix. It possesses the qualities of laurium and iceberg blue calcite, both of which allow more light to pass through their matrix. Therefore, cuprian aragonite is a power behind the throne mineral which resonates to the Virgo ability to serve behind the scenes without seeking acclaim or limelight. It resonates to the advisory role and enforces the awareness of making only statements which can substantiated. It is a stone which repels gossip and heresay, and is excellent for the judge's chambers as well as the courtroom and the critic's office. It tends to soften the self-critical nature of Virgo, and smoothes comments made by the critic so they are well-received. See Blue/Gray Danburite. *(5th chakra)*

Golden - The interesting quality of this mineral is that it shows the spherical nature of being in light while it also expresses the aspects of self reaching out to experience life, knowledge, and physical essence. It is an empowerment mineral in the past given to young adults in recognition of the responsibility of adult tasks now being required of them. It possesses the ability to strengthen the internal chakra chord so that it is able to bear more light and therefore more "witness" of the higher self. It is a subdued essence of mineral allowing great tranquility and self-possession of personal power which may be required of young adults so that their competitive drive can be expressed more toward enthusiastic cooperation. The spherical nature of beingness is what allowed Columbus to first deduce that circumnavigation of the globe was possible. The quality of committing oneself to the proof and attainment of spiritual spherical essence is supported by aragonite. The knowledge of self as a bundle of vibratory frequency in molecular matter is supported by aragonite as well. *(3rd chakra)*

Purple - This form of aragonite is found in Spain in flashlight battery-size single crystals. This tone completes the awareness and nature of being in light by bridging to spiral energies of DNA, the entwined serpents of the caduceus, kundalini flow up and down the spine (nadi and sushumna), and entwined ley lines. This is also symbolic of the labyrinth, the vortex, the spinning and whirling ritual dance of the Sufi, and the spiral galaxies of the universe including our own Milky Way. It is a pure revelation of mineral consciousness to see how the Spanish flamenco dance honors the vital power of this mineral. Purple aragonite fulfills the urge to harmonize with these patterns of evolution of energy and matter. It brings consciousness deep within, yet spins it ever outward. This vibration is the original teacher and conduit to living organisms of the spiral. It holds the dual movements of death and rebirth, in kalpa and pralaya, and it is the division line between yin and yang. It is the informing principle of how to return home after death. The energy of purple aragonite resonates to grouse medicine. *(All chakras)*

White - These wonderful stalagtites and groups of florets occur in clear to creamy white shades. The many paths of enlightenment converge upon a single truth: all or nothing at all. All is vibration. Nothing at all is the Void. From the Void came all. When we arrive at the place we knew all along, there is one choice: to remain in vibration or to return to silence. Aragonite brings to the body a cessation of noise which allows momentary and near perfect awareness of order. All is known. Nothing at all is the unknown. *(7th chakra and above)*

ARFVEDSONITE: Arfvedsonite is a black amphibole-group mineral forming in short, flat pencils. It is a hydrous double sodium double iron silicate classified in the amphibole group. It is more absorptive than schorl or obsidian, therefore it is superior for drawing negative energy off the base of the spine and first chakra areas. It is cleansing, protective and grounding. Arfvedsonite can do a great deal on its own to provide stabilization, endurance, stamina, and foundation, therefore an individual isn't likely to need additional backbone supports in these areas. Arfvedsonite is a good choice to remove residual anger and any issues of resentment or bitterness which remain about loss. It aids the individual to let go of feelings of victimization and negative relationships to loss. For individuals who have a hard time saying no to others or establishing boundaries, arfvedsonite is an excellent mineral support. It is recommended for anemia, spinabifida, and paranoia. Its vibration is highly masculine. See Riebeckite. *(1st chakra)*

ARSENOPYRITE: Arsenopyrite is an iron arsenic sulfide. It occurs in silvery, metallic crystals which can be pyramidal or tabular. It is the "shop steward" of minerals, because it does a good job of supporting those who represent or speak for the rights of others, giving them confidence to verbalize and explain their forthright values. It enhances "right responsibility." It is a mineral which simplifies conflict and deepens the bonds between people. Arsenopyrite cuts to the quick and to the bone with its simple, clear energy, and brooks no dissent which is based in fear, complaining, victimization or violence. *(1st chakra)*

ARTINITE: Artinite is a hydrous magnesium carbonate. It looks like a white fuzz ball, and is known as a thinker's stone, because it is especially attuned to mathematicians, astronomers and those trained in physics. Its fine filaments carry light frequencies of such incredible clarity that streams of complicated numbers and thought processes flow easily. It adapts well to highly technological environments. *(7th chakra)*

ASTROPHYLLITE: Astrophyllite is a hydrous silicate containing potassium, sodium, iron, manganese and titanium. The long fans of golden needles in acicular pattern draw you into the mineral. Found in granitic host rock, this mineral improves attention span, concentration, and focus. It strengthens short term memory, and enhances attention to detail. Its micaceous structure strengthens making sense of many pieces of information, therefore it is useful to those who investigate and research. *(1st and 7th chakras)*

ATACAMITE: "The Alchemist's Stone." Atacamite is a hydrous copper chloride usually forming in green crystals laying down like grass on a green background. The nicest specimens are found as transparent green icicles in Chile. Since atacamite fosters maturation, it is an excellent stone for Capricorn, Aquarius, Pisces and Aries (in feminine astrology, Aries is the last sign of the Zodiac) individuals. It supports the mature and aged. It is a stone of development to peak and fruition of ideas, projects or living things. It supports following formulae or recipes to completion. For a better idea of how to utilize this mineral, refer to the "alchemy" entry in the Properties to Minerals Index, which lists the minerals with alchemical properties. Since salt (sodium chloride) is one of three stones used in alchemical work, the chlorine in atacamite provides a necessary physical connection for the copper to transmit life force energy. Alchemists had to follow a meticulous and rigorous step-by-step regime which was the beginning of modern day chemistry and chemical technique. It was not uncommon for alchemists to devote their life's work to perfecting their technique. Undoubtedly, atacamite provides the foundation for patience, wisdom, maturity and "practice makes perfect." *(4th chakra)*

AUGITE: Augite is a member of the pyroxene group of minerals, which includes jadeite, spodumene and diopside. Augite forms in black, blocky crystals often associated in granite or mixes of minerals, such as with peach calcite from Ontario, Canada or in charoite from Russia. It is a powerful grounding, mandalic form which benefits altars and areas of sacred and creative activity by its presence. It is a mystic mineral with deep value to seers, mystics and spiritual leaders. It is an excellent aid during times of spiritual crisis and questioning, especially brought on by shock, crisis, loss, grief and death of a loved one. It is suggested especially for the early stages of doubt, as it provides a background of trust of return, of withstanding and honoring the "dark night of the soul." It is recommended for soul retrieval work. *(1st chakra)*

AURICHALCITE: Aurichalcite is a hydrous zinc-based copper mineral, and forms in bright, fuzzy coatings which occur in Mapimi, Mexico. It is the yogi of copper minerals. It is especially applicable in meditative and co-meditative practices. It is calming, much more so than amethyst. This is because aurichalcite calms the mind circuitry. The mind circuitry is composed of the brain and the nervous system separate from the chakra chord. The chakra chord is calmed by amethyst, and a tube of purple light funnels down the human body through the chakra chord. The reason brain tissue and spinal tissue are so sensitive to injury is because of their molecular structure which accommodates the low frequency purple light.

Aurichalcite often allows an individual to overcome paralysis or seizures by recircuiting disrupted channels of physical neurons to new pathways in the brain and spine. Recircuiting is essentially an easy principle to manifest once the basics are understood. Aurichalcite is a conduit which allows that basic information to be provided to the greater self where it can be enacted on the physical portion. Injury to the spine can often be redigitized or healed through the use of large amethyst crystals with aurichalcite. Aurichalcite is a cooling mineral, useful for temper disturbances and heat stroke. It assists in balancing body electromagnetics. Aurichalcite is found in the Etheric Conditioners category of the Spiritual Apothecary. *(6th chakra)*

AUTUNITE: Autunite is a hydrated calcium uranium phosphate which is radioactive. It behaves like an antenna to aid specific focus to energy forms. It can also behave like a transmitter, sending clear, unmistakable signals to distant places. It is an excellent mineral for people who channel highly technical information because it increases language accuracy between the different languages of the universe, be they spoken or not. Autunite increases "wrap" awareness where all incoming stimuli to a frequency are blocked so that incredible breadth is achieved across a continuum of that frequency without disturbance. This strengthens the ability to send healing long distance. Autunite is typically yellow with lime green edges. *(3rd and 4th chakras)*

AVENTURINE: Aventurine is a common name for a form of massive quartz with abundant chromian mica and occasionally layers of pyrite or hematite inclusions. It is often visually arresting and physically energizing. All colors of aventurine are normally found in India, and just recently a light blue and lavender color from Brazil.

Blue - Blue aventurine is second only to clear danburite in supporting masculine shamanic energies. It can be used with danburite for superior results. It opens the third eye. Blue aventurine is usually given to an initiate before danburite, because the aventurine prepares the initiate for the powerful awareness stimulated by danburite. *(6th chakra)*

Green - Green aventurine carries the healing power of the green earth. It heals the heart by releasing emotional stress, thus allowing reuniting of the healthy heart and soul. It has shiny silver or golden pyrite flecks, commonly, which enhance energy flow to other body areas. *(4th chakra)*

Lavender - Or "sampsonite." (Ha! Ha! The luggage stone. Pack up your troubles. Let's get happy!) This stone is a real treasure for those suffering from depression. It supplies more energy, a decreased level of worry, and a sense of serenity about life's experiences. The "sampsonite" quality is not supposed to allude to luggage at all. It relates to the strength of Sampson and spiritual conviction in one's path and actions. This is an excellent stone for spiritual advisors such as priests, nuns, reverends, counselors, psychics, etc. *(6th and 7th chakras)*

Peach - The "whisper stone." This mineral is best used in quiet mediation because its qualities are best shown when the individual is seeking centering and quiet. It is a good mineral for shy people and people full of anxiety. For shy people, it provides a sense of dignity and presence. For people full of anxiety, it reminds them to calm themselves to listen to the heartbeat of life and know all is well. It helps to still the loud, critical voice within. Actually an orange/brown color. *(2nd chakra)*

Red - A genital and reproductive organ healer, red aventurine can be used effectively to reverse disease states in these areas. For cancer, red aventurine must be complemented by bloodstone, abalone, paua shell and one or all of three green minerals: malachite, dioptase and green garnet. Red aventurine is an excellent mineral to counteract too much air in the zodiac, because it resonates to earth, fire and water. Because of this, red aventurine must be used with minerals which oxygenate the blood, as listed above. *(1st chakra)*

AXINITE: The most common mineral in the axinite group is ferro-axinite, which is a borosilicate, meaning it contains boron. A glassy brown mineral, axinite is as excellent a grounder as black tourmaline. Axinite is a feminine energy counterpart to the masculine energy of schorl (black tourmaline). Axinite stimulates body awareness of energy fields, and enhances neurotransmitting functions, as well as the transmission of healing energy through the hands. It is an excellent support to dowsers and those sensitive to waves of energy or changes in p, r, and s waves prior to an earthquake. As a member of the boron-containing minerals, axinite possesses shamanic properties which transmit to humans as strengthened ability to handle immense influxes of energy. *(1st chakra)*

AZEZTULITE: Azeztulite is a casual name for an excellent seventh and above chakra stone, which minimizes annoying mental activity during meditation or psychic work. It should not be worn at any other time as it decreases motor skills. It should not be worn in an unprotected situation or environment, because it has affinities to Azazel when improperly used. Azazel, in Hebrew tradition, was the rebel leader of the angels who seduced mankind. Apparently, used in conjunction with extraterrestrial angelic beings known as the Azez, Azeztulite helps the planet and people to transition to higher frequencies, and the stone is said to be their special conduit for that energy. *(7th chakra and up)*

AZULICITE: Azulicite is a gemmier variety of sanidine. The moonstone-like flash qualities of azulicite create a translucent blue shine on one end of the gem. When inserted between fingers (four stones) and toes (four stones), it is an excellent meridian cleanser and balancer for the entire body. The user places the eight stones in position and relaxes to allow the work to progress. The hands can be moved into sacred positions to enhance the color vibration in the healing. Tingling and itching indicates the meridian is releasing. Five- to 10-minute applications on both hands and feet are excellent. This procedure is especially useful for practitioners using their hands in healing work, as it allows the individual to resume balance quickly after taxing sessions. *(Hands and feet, also 6th and 7th chakras)*

AZURITE: Azurite is a hydrous copper carbonate which can occur in a blend with malachite, in round balls of crystal, or as tabular crystals which are stunning in their cobalt tone and occasional clarity. Azurite's rich, dark blue color stimulates psychic powers and inspirations, and improves visuals in meditation and intuitive work. It aids consciousness development, mental processes, and connection to the higher self. It activates the spleen and thyroid while opening the etheric body to the highest protection color, cobalt blue. Azurite, as a copper mineral, bears the structure to empower protection of the life force in the physical body. It is found in the Etheric Opener division of the Spiritual Apothecary. Azurite has historically been highly regarded for the depth of its cobalt blue color. Azurite is unequaled as a feminine power mineral for the third eye. It stimulates prophetic awareness, often being called spiritually "Cassandra's stone." When using this mineral consciously, one must be willing to take on the responsibility of being the messenger. Azurite increases the physical healing effectiveness of iron-based minerals, such as quartz with a yellow stain on the outside, pyrite, hematite, red jasper, goethite, citrine quartz, etc., because copper stimulates the assimilation of iron. *(6th chakra)*

AZURITE/MALACHITE: The healing green of malachite softens azurite's powerful and intense blue energy. This natural mineral combination activates and heals congested or inflamed areas of all organs, bones and tissues, provides nervous system comfort, and reduces stress. It aids in healing liver and skin tissue. It blends the energies of the heart, throat and third eye, therefore it is compatible with amazonite. It is literally an intercessory mineral between electrical life force impulses and the earth and living organisms. Its ability to express itself visually as the earth does when viewed in space speaks to its embodiment of the life force principle for us here on earth. Azurite/malachite is located in the Spiritual Apothecary in the Etheric Opener division. Try this combination for supporting the electromagnetic field of the body. *(4th, 5th and 6th chakras)*

BARBERTONITE and MAGNESITE: Near Zeehan, Tasmania, we dug in a small hill where this mineral is found. Barbertonite is a magnesium chromium carbonate which is dimorphic with stichtite. Barbertonite is a connector to the constellation Pleiades. It has more green aspect (serpentine) than stichtite which emphasizes the "Earth" end of the connection and the heart healing necessary for humans to ascend to purple as a mid-spectrum color at the seventh chakra, rather than Earth's green at the fourth chakra. The seventh chakra sits midway between the twelfth and first chakras, and represents in consciousness evolution the true half-way mark to realizing adeptship. See Stichtite. *(4th and 6th chakras)*

BARITE: Barite is barium sulfate, and is categorized in the barite group with anglesite and celestite. Barium is a very heavy element, so barite has a pleasing, weighty heft to it. All members of the barite group transmit and uphold frequencies of truth, honesty, and integrity. Barite is calming to people who feel too mentally or physically active when it is placed in their living/working areas. It increases harmonious influences in interpersonal relationships as it helps to release hostility or resentment in a healthy manner, and it supports communication links. All forms of barite soothe tissue of the digestive tract and increase the excretion of bile. A popular form is Desert Rose. Both the white/clear and blue forms are found in the Spiritual Apothecary in the Auric Conditioners division. *(7th chakra)*

Blue - Blue barite is usually found in small, clear, flat, chisel-tip crystals with beveled terminations. Historically, blue barite has been extremely fashionable, like diamonds or emeralds. The power of blue barite is its unequaled ability to calm and tone the being on many levels during trauma and after extreme shock. In the past, it was used widely by women attending men after a battle. When placed on the third eye of an individual, it often allowed a semi-transcendant state to occur, so the individual could survive waiting for treatment if not attended to immediately. These women achieved such success in saving lives that many people ascribed magical qualities of immortality to barite. This is not the case, of course, however the tranquility ray of blue barite is exceptional and highly evolved. See Benitoite. *(6th chakra)*

Clear - Clear barite usually forms in small, square crystals. Like clear quartz, it is adaptable to multifaceted use. It is both a body energizer and astral body cleanser. It creates current flow between aspects of the total self and is an optimizer in psychic work. In meditation, it polarizes information to enhance clarity so that, in emotional issues, a more accurate reading of one's own intent can be deduced. This is true for reading others as well. *(7th and 12th chakras)*

Gold - Gold barite often occurs as blocky tabular or rectangular crystals which can be clear or opaque. They can form as single crystals or in prismatic clusters. This is the "golden seal" and was often prized by monarchs to be sent along with state documents as a sign that the transaction of documents met the highest standards of truth, dignity, and honesty. It represents the highest truth in written form. If these aspects were not being followed, sending the stone in dishonesty, etc., meant as much as a death warrant to the monarch exposing himself thus. The golden seal is just as applicable in modern times as an indication of truthful intent in one's work, and can be used in meditation to help clarify responses received from the other realms to questions of extreme import to the individual. This is an excellent stone for writers, detectives, lawyers, and all people who represent others such as statesmen and political leaders. Golden barite is the embodiment of the clan chief, cougar, puma, and mountain lion. See Anglesite and Celestite. *(Upper chakras)*

Rose - Barite roses, also known as desert roses, are heavy-for-their-size concretions of barite with sand. The barite forms in small discs from which other discs protrude. They can be more rounded or more angular and disjointed, depending on the specimen. The addition of sand (or silica) to the barium sulfate makes barite rose an effective complement to skin and internal cleansing regimens. It supports shiny hair and smooth complexion, as well as strong nails, teeth and bones. It helps to counteract loose skin and chronic fatigue. This is a good combination of elements to counteract tuberculosis. It is excellent for bile excretion and bowel cleansing. *(1st and 2nd chakras)*

BARTHITE: Barthite is a green, fuzzy malachite-type color mineral which forms a coating on its host rock. It is quite rare. Barthite aids in developing consensus. It helps to conceptualize and capsulize thoughts or theories which are still in formation. It draws out the inherent essence of ideas making them easier to explain. *(7th chakra)*

BENITOITE: Benitoite is a barium titanium silicate. It is the state gemstone of California, where it is found in San Benito County. It occurs in small, triangle-shaped crystals with beveled edges, something like the shape of a "yield" sign. It bears a deep cobalt blue color, and often is found associated with neptunite which is also dark blue. The blue/purple varieties (some have white bands) of benitoite speak for its highest level of activation powers for creativity and self-awareness. It strengthens clairvoyance and visualization during psychic work or spiritual practices. Benitoite possesses other-worldly properties due to titanium in its chemical composition, therefore it is excellent for spiritual guidance and making stronger interdimensional contact. This is a fluorescent mineral which allows for insight through darkness.

Benitoite is a higher order presence of blue barite. Benitoite was and can be used as an activation mineral for long distance healing support when healers use blue barite in the field. Also, benitoite was and can be placed with a generator quartz programmed to provide greater levels of energy for spiritual rescue work in the field. This was done on battlefields where many casualties resulted in death, and the transfer of souls all at one time required huge influxes of loving support to see them on their way "home."

It is not unusual for benitoite to enhance gentleness and mental genius in those who appreciate it. It also strengthens the ability to bring in technological and mechanical information, so it is useful to inventors, futurists and computer gurus. It harmonizes well with kundalini flow already present in geniuses, psychics and creative people, and contributes to increased kundalini flow. *(6th chakra)*

BERTRANDITE: Bertrandite is found in an opalite base, and it forms in nodules and seams. It is a hydrous beryllium silicate. Some old-timers call it beryllium fluorite. Formed into cabochons, bertrandite can range from deep purple with black lines to the same intermixed with creamy white and lavender. This combination absorbs negativity while transmitting calming, soothing energy. Bertrandite is good to use during or after physically invasive work such as dentistry, first aid or triage. It supports the nerves to transfer and release shock and discomfort out of the body, and greatly reduces the effects of trauma. Its violet/ purple swirls in creamy white background provide a visual stimulus to release traumatic memories, and is especially applicable to reversing the abuses of mind control, psychological torture, programming, brain washing, kidnapping, and abduction. *(6th and 7th chakras)*

BERYL, Common: The transparent varieties are aquamarine (blue), emerald (green), heliodor (yellow), and morganite (pink), but the white, gray, or pale blue ones are opaque and termed "common." Beryl is a ring silicate, where stacks of molecules one above the other are held together by beryllium atoms. Beryllium compounds are quite toxic. Berylliosis is a painful and fatal disease resulting from exposure to and inhalation of finely powdered beryllium compounds. All beryl minerals are found in the Spiritual Apothecary, the only mineral group to be specifically utilized in five of the nine categories. This is because the toxicity of beryllium empowers beryls to transmute from lower to higher forms. They are naturally homeopathic.

Beryls represent serpent energy released in the body, and it is not uncommon for organisms to require time to handle their energy. All beryls increase kundalini energy in the body for healing as well as consciousness expansion and development. It is commonly believed in many cultures that life on earth is impossible without Serpent. This is likely to be true, because movement of energy up and down the spine is a kundalini process. All consciousness development is dependent on kundalini. Yet, Serpent introduces a toxicity which must be transmuted, hence the image of serpent has become a complex, often negative symbol for humans to decipher. Common beryl stimulates the root chakra, providing strengthening for males and cleansing for females, and can be used for all lower digestive disorders, including bowel cancer. Beryl was used historically by scryers rather than clear quartz crystal, and it aids in preventing a psychic from being used or manipulated. Its use is strengthened by the moon. It is a mineral of mediation, friendly negotiation, and friendship. Beryl is the state mineral of New Hampshire. *(1st chakra)*

Blue - See Aquamarine.

Clear - See Goshenite.

Green - See Emerald.

Pink - See Morganite.

Red - This small, red hexagonal crystal is found only in Utah. Red beryl is a mineral of self-knowing. Often called the mirror stone, as sensations of waving, reverberating light lines lead the individual to a central core which is immediately recognized as the true self. This core is the vibratory link to the higher self where all guidance and inspiration emanates. In the Spiritual Apothecary, red beryl is found with emerald in the Etheric Sealer division. *(1st through etheric chakras)*

Yellow - See Heliodor.

BETAFITE: Like a good rain, betafite is a cleanser of auric and etheric bodies. It can be passed over the body after other cleansers are used to pick up trace elements and elementals, which are the seed hosts for attraction of whole communities of unwanted energies. Betafite is good to use before psychic surgery and during subconscious interactions such as hypnosis and soul retrieval. It provides a form of psychic protection through clearing subtle body layers, however other minerals are recommended for protection. It is appropriate as a companion mineral for all out-of-body work, such as dreamtime and journeying, as it releases unwanted energies from the soul force prior to re-entrance to the body. Betafite is found in both the Etheric and Auric Cleanser categories of the Spiritual Apothecary. *(Auric and etheric chakras)*

BIOTITE (Mica): "Scout Stone or Observer Rock." Biotite is a form of brown mica which enhances tracking and scouting abilities as it stimulates visual receptors on the physical and psychic levels. It associates well with machines, and boosts the ability to observe how well a machine is operating. Biotite is a series name for micas without lithium. The chemical formula transitions from potassium-iron-aluminum to potassium-magnesium-aluminum. It is difficult to know whether a specific specimen is iron-rich or magnesium-rich, but the energetic difference will express itself for identification as either energizing (iron) or soothing (magnesium). Iron-rich specimens oxygenate the body and serve blood and respiratory functions. This type will aid in thinking clearly and strengthening the heart and liver. Magnesium-rich specimens aid nerve fibers in sending messages to muscles and bones. It relieves muscle tension and involuntary contractions. It assists glandular function, especially for sterility and thyroid problems, and stimulates enzyme production. It is a good mineral to reduce stress. *(All chakras)*

BISMUTH: Bismuth is an element which is a semi-metal. It is a member of the arsenic group along with the elements antimony and arsenic. It is a silver-colored heavy metal in nature which cannot express its crystalline structure without recrystallization help from humans. Bismuth does not crystallize naturally, because it is usually too crowded by host rock to form as a crystal. When expressing itself in recrystallized rainbow iridescence, it is excellent for focus, visualization, and shamanic journeying. It aids in shapeshifting. Copernicus would have loved having a study lined with bismuth, because it lifts awareness to the stars just like an elevator with a palpable sensation of journey. What is most clear about bismuth is how far it can go into the universe. It provides a solid sense of grounding while this occurs. *(6th chakra and others, depending on colors)*

BIXBYITE: A magnesium oxide with the appearance and affect of hematite (iron oxide). It forms in cubes. Bixbyite grounds and energizes like most metals, and protection is offered by its reflective quality. When associated with golden topaz, it provides a powerful grounded connection to the upper chakras and opens the circuits of will at the same time. *(1st chakra to upper)*

BLOODSTONE: Bloodstone is a specific form of chalcedony quartz tending toward a jasper, found most readily in India, with red spots (caused by iron oxides) imbued into the green stone. A recent find from Oregon is quite beautiful. The Biblical reference refers to the symbolism of Christ's blood spilling on stone. In ancient texts, it is often referred to as heliotrope. Bloodstone from Oregon is quite powerful and possesses a smooth quality of ease and gentleness. Bloodstone grounds the energetic body to the physical, and cleanses the physical body helping with all blood disorders. It strengthens physical balance and increases courage as it resonates powerfully with the heart chakra, allowing blood oxygenation to increase. It supports the role of the heart as an organ of awareness and learning which can adjust itself to changing circumstances as the result of experience. It is a stone of surrender, humility and recognition of one's service to others, hence a deeper connection to the story of Christ and Buddha emerges as well as the knowledge that compassion is the only way out of suffering, whatever one's spiritual persuasion. Bloodstone is the March birthstone. *(1st and 4th chakras)*

BOLEITE: "The Surgeon's Stone." Boleite is a hydrated lead copper chloride also containing silver. It is found historically as small, blue cubic crystals in Baja, Mexico. Boleite promotes steady muscle movement, efficient firing of synapses and precise mind/eye/hand coordination. Called the "Surgeon's Stone," Boleite is renowned in the spirit realm as a psychic scalpel, because it is so effective in releasing unwanted ties, energies and entities. This translates over into the physical as superior surgical skills especially in removing tumors from sensitive areas. Boleite is extremely powerful in stimulating manual dexterity in children before the age of six. Energetically, boleite cauterizes and cleanses the area where psychic surgery occurred. Boleite is a Healer in the Spiritual Apothecary. *(Hands, 5th and 6th chakras)*

BOULANGERITE: Boulangerite is a lead antimony sulfide. Its hair-like, silver fibers usually lay down in masses. It is a subtle body mineral often too refined to be of much use to modern man because of the toxin load carried in the body. However, it is beneficial to use cautiously during cleansing therapies and afterwards for a few weeks, because the clarity of the body will allow the antimony stimulus of boulangerite to operate optimally. During these times, life visions can occur which provide the individual with guidance regarding life purpose and mission. Good to use during times of spiritual seeking and surrender.

BRASS: Brass is a copper and zinc alloy with deep healing properties for the physical body and immune system. Copper associations to solar energy, the fire element and the planet Venus empower it to act for divine love, chivalric love, enthusiasm and physical grace. It can be thought of as love in action, being an example of love rather than just a spokesperson for love. The word *brazen* derives from brass, and has come to an unfortunate negative connotation, such as "brazen women" being a term for strong women unconcerned about judgements about their sensuality or physical agility. Perhaps its complimentary meaning can regain favor. It assists hair growth and skin rejuvenation as well as protection of the skin layer by penetration of bacteria and viruses. It can be worn as a breastplate and/or backplate as a form of energetic protection. Brass fittings on water systems tend to transmit the benefits of brass as an energetic agent, however it can be more consciously employed by placing it in living environments. *(All chakras)*

BRAZILIANITE: "Messenger stone." Brazilianite is a hydrous sodium aluminum phosphate. Found in its principle locality in Brazil in 1940, this mineral is olive green/yellow. It forms in blocky, prismatic spear-shaped crystals. Brazilianite complements the work of aligned individuals in separate kingdoms such as hawk and peyote, both aspects of messenger power. It bears energies of the planet and god Mercury. It is a highly regarded mineral on the spirit plane for its ability to stimulate the psychic abilities of the third chakra. Because it is found naturally associated with blue apatite and lazulite, the three work together to enhance verbal psychic work such mediumship and trance channeling. *(3rd, 4th and 5th chakras)*

BROCHANTITE: Brochantite is a hydrous copper sulfate. Brochantite attracts and draws in high universal energies. It is an "Earth Keeper" stone, meaning that it has strong Earth resonance and is tuned to Earth system needs and balances. It is best used with pyramid energies as the radiation from the combination is exceptionally healing and empowering. In healing wands, brochantite can be placed at the end where the hand rests or in the middle. Crystals at the transmission end will release only the highest of healing vibrations. *(4th chakra)*

BRONZE: Bronze is an alloy of copper, tin and silver. It combines masculine, fire energies with feminine, water energies. Copper possesses association to solar energies as well as the planet Venus and silver is a lunar/Moon element. It provides the voice of love, the harmonic grace of high and low tones. Bronze is one of the symbols of justice. Bronze has been used worldwide for spiritual objects, including candle sticks, bells, bowls, incense braziers, and chalices. Bronze was used for breastplates, and was believed to confer immortality as well as incorruptibility to warriors and the spiritual alike. It carries vibration deep into the bones, blood, and cell level, thereby restoring balance. Its graceful shine and tone promotes harmonic inner rhythms necessary for healing and alignment. It can be thought of as the alloy which provides sound acupuncture. *(All chakras)*

BRONZITE (Massive brown metallic): Beautifully chatoyant from red to golden patches like brush strokes. Highly tuned to musical personalities or atmospheres promoting benefits of musical interactions. Stimulates self expression and physical grace. Provides strength to withstand discordant atmospheres or frequencies such as court rooms, hospital emergency rooms, police stations, etc. Enhances harmony. *(1st and 2nd chakras)*

BROOKITE: Brookite is simple titanium oxide. It forms in black, small, flat and square crystals. It is found in Magnet Cove, Arkansas. Brookite is a good mineral to use when bringing closure to a gathering or project. It can be used at the end of a relationship, but with mixed results as some people are not quite ready for the firm finality this mineral shows. It enhances the awareness of time and meeting a deadline on schedule. It allows people to feel complete at the end of an event. *(1st chakra)*

BRUCITE: Brucite is a simple and massive magnesium hydroxide. The nicest white/ blue chunks occur in Italy. Brucite supports all bodily functions which utilize magnesium including bones, nerve fibers, glands, muscles, teeth, enzymes and nerve impulses. It is excellent to counteract the stress of modern life, alcohol consumption, neuralgia, and thyroid problems. It is superb as a toned or tonal mediation tool where the stone is held in the left hand while the individual releases tones from as deep in the chest cavity and abdomen as can be reached. Excellent for opening up the entire breathing system especially for asthma and emphysema, and before singing, oratory or exercise. *(5th chakra)*

BUDDSTONE: An African black, green and white mixed stone, related to verdite, buddstone is a connector for the first, fourth and seventh chakras. Buddstone aids all forms of energy movement in the body whether it is kundalini, psychic or nutritional. As a broad spectrum stone, buddstone empowers many types of individuals and activities, from childhood to old age. It can be especially soothing to individuals with Alzheimer's disease or senility. *(1st, 4th and 7th chakras)*

BURNITE: Burnite looks like azurite. It alleviates symptoms of dizziness, nausea, motion sickness, and vertigo, and in general, has application to inner ear problems. It is also beneficial in reducing distress from earaches and ear infections. Being rich blue and green, it looks like azurite/malachite. *(4th and 6th chakras)*

BUSTAMITE: Like a shoreline, bustamite represents a place of contact diffusion where two energies are rubbing against each other. This interaction enhances flexibility and adaptability for those who find themselves rigid and unable to change or leave behind past survival habits or judgments. For people who are already quite flexible, bustamite may be too stimulating, and should be avoided for conscious, daily use; however, it performs excellently as a meditation deepener.

Bustamite enhances what can be called "edge consciousness or behavior." As theorized by Cook, edgelings are people who experienced child abuse and/or alien abduction and/or exhibit psychic abilities. Edgelings adapted or utilized out-of-body escape to survive abuse or experience out-of-body states frequently due to psychic events. Cook says edgelings indicate where civilization is going, and they provide important examples to others of how to integrate this awareness and function in daily life. In biology, environmental edges support the densest populations of species, a good indication that contact zones provide the richest and most diverse resources and opportunities. Bustamite is found in the contact zone between manganese and sugilite. Bustamite is a powerful ally to assist shifting paradigms, and is applicable to cultural as well as personal levels of changing belief patterns. Chemically, bustamite is similar to rhodonite, and it does not contain lithium as sugilite does. *(1st and 2nd chakras)*

CACOXENITE: Cacoxenite is a hydrated iron aluminum phosphate, which rarely forms minute crystals. For use as a mineral, cacoxenite is found most easily as golden yellow needles in amethyst crystals, usually from Brazil. It harmonizes well with the electromagnetic field of the human body, and enhances psychic processes. Cacoxenite is an energizer that impacts metabolism of protein, carbohydrate and fat, increasing the availability of all. It increases mental alertness through assisting the production of lecithin and cerebrin. It supports all glands and the kidneys. Combined with amethyst, cacoxenite increases alertness in sleep for lucid dreaming and in meditation for clairaudience and hearing the still, small voice within.

Cacoxenite works optimally in combination with calcite, danburite, and gypsum (selenite). The calcite combination works well on physical issues related to nerves exhausted by the demands of daily life. Therapeutically, this combination increases circulation, and helps to increase sluggish heart function. It elevates low blood pressure allowing sluggish people to function at their peak. With danburite, cacoxenite restores depleted energy in people who devote long hours to stressful caregiving of others, such as healing practitioners, hospital and emergency room workers, dental workers, etc. With gypsum (selenite), cacoxenite replenishes nerve centers exhausted by intuitive and counseling work, such as mediumship and therapy. *(All chakras)*

CALCITE: Calcite is calcium carbonate, the basic building block of skeletons and exoskeletons which we know as shells. Calcite can form in literally hundreds of crystalline structures, an unknown capacity in any other type of mineral. It is also a mineral which rounds naturally when it weathers leading to the rounded character of living organisms. The calcite family expresses itself in many colors from clear and white through deep orange and brown, usually in massive form. Clear rhombohedrons exhibit a light characteristic called double refraction.

Calcium and calcite minerals hold a feminine vibration, and calcite is the ombudstone from minerals to life. Just as crows are want-to-be hawks, many minerals want-to-be calcite. That is because calcite alone represents the central axis round which all else revolves. It can be thought of as the umbilical cord of minerals (see orange calcite, below), for it is the staff which connects spirit to life through minerals. In the forms of marble and limestone, calcite achieves equally remarkable capabilities. All forms of calcite are etheric cleansers, and they are located in that category in the Spiritual Apothecary. All forms of calcite (including limestone and especially marble) facilitate and respond to sound. The use of marble or limestone as a temple or sacred structure building material is important in facilitating altered states of consciousness, healing, the maintenance of life force over time, shapeshifting and time travel. Chanting, toning and singing in these structures creates the frequencies necessary for all these manifestations to occur. Stan Tenen, a mathematician who studies patterns in the Hebrew text of Genesis, says the Great Pyramid of Giza was such a structure, and at one time, the limestone blocks visible today were capped with marble which was removed to rebuild the mosques of Cairo after earthquakes. He believes a smaller version of the Great Pyramid could be built to test his hypothesis that the pyramid is a shapeshifting and time travel device activated by sound.

Blue - Blue calcite worn on the body is a calmer, and is effective as a sedative after emotional trauma such as the death of a loved one. It is located in the Etheric Openers category of the Spiritual Apothecary. It is a powerul aid to drug (anesthesia as well as pharmaceuticals and addictive), chemotherapy and radiation detoxification.Taken as an elixir, it is a powerful aid to the colon and liver. One is advised to place the calcite specimen in a glass separate from the elixir solution, and activate in sunlight for four hours. The colon and liver are cleansed by blue calcite, and it can be worn as support during a period of treatment to these areas. It is suggested that 15 drops be followed five minutes later by at least eight ounces of water and no more than 20. This will allow the mineral to alert the body's own cleansing system to release enzymes in the stomach which are carried through the digestive tract to the colon, and absorbed by the bloodstream to be transported to the liver. Take 15 drops in the morning before eating. Eating can occur 20 to 25 minutes later, and take again just before bedtime. If diarrhea results, move the evening dose to the afternoon when the stomach is empty; especially important to follow a meal free of complex protein. Continue to drink water every hour (eight ounces).

This treatment can be used for two weeks to induce deep cleansing. Afterwards, maintenance can be achieved by using treatments one or two weekends a month, or one day a week depending on the severity of the case. The physical abnormality leading to requiring blue calcite is lack of enzyme production relating to organ and tissue regeneration in these areas. Chronic constipation over a lifetime shows this enzyme deficiency, but this treatment also is excellent for less chronic colon and liver ailments. *(1st through 7th chakras)*

Green - Green calcite is called "the joy stone," as its tantalizing vibration feels immediately pleasing. It aids heart and lung issues. This is definitely a stone for hard workers, especially those who do jobs of drudgery for others, such as house cleaners, maids, dish washers, garbage collectors, etc. It is a stone which brings in the Buddhist regard for all activities of man being equal in importance and stature, especially when done with grace and a smile. It is a mineral of the ecstatic state, and is a perfect exemplar of the injunction, "Before enlightenment, chop wood, carry water. After enlightenment, chop wood, carry water." *(4th chakra)*

Honey - Honey calcite is a golden color which connects energies systemically in all the bodies, physical, etheric, auric, etc. It activates root chakra connection to the broad spectrum of universal energies and concepts which support the ability to act as a humanitarian and advocate for the disadvantaged in ways which support and honor the highest good of all involved. It encourages global awareness while focusing on being able to act in one's own spehere of influence, and enhances the ability to share skills and talents in a group setting while maintaining integrity of self with healthy and appropriate boundaries. It stimulates kidneys to remove body toxins as listed above, and also stimulates action in lethargic or sluggish people. *(All chakras, primarily 1st and 11th)*

Iceberg Blue (Bisby, AZ) - This unusual tone of seafoam blue calcite can be layered in bands of pearly white and blue or some specimens can be dispersed throughout with blue texture. Related to laurium or light blue aragonite, uniformly blue calcite is the "voice of authority." It showers the lower brain with ancient memory activation, so it can be considered a voice of experience as well. The trace of copper in iceberg blue calcite makes it energetically stimulating to the brain. It encourages curiosity, inquiry and investigation as well as the ability to interview others, delegation of authority, and giving orders. Layered specimens of iceberg blue calcite link the fifth and seventh chakras for a "Sermon on the Mount" or "I Have a Dream" type of oral power. The entire skull becomes a resonance chamber for divine frequencies of sound. *(5th and 7th chakras)*

Lime Green - Lime green calcite bears the vibration of "Nothing changes. Nothing stays the same." In the urge to translate light energy and information into mobile, living organisms, lime green calcite raised its molecular signature spin to the required level to cheer on the new while holding steady the old. It is the ultimate cross-over mineral in the calcite family which defines making a leap of faith into the unknown quantities of plants and animals. *(3rd and 4th chakras)*

Orange - "The Midwife Stone." This color and quality of calcite supports life force, and is a powerful ally for any person or practitioner who mentors, assists, aids or empowers another. It is the ombudstone which connects the seen and unseen elements of spirit and matter. Many are called to be midwives at the birth of new life, projects, ideas, ages or causes, and the vibration of orange calcite is creative, willful and expressive, providing empowerment to those who can benefit from it. Because orange calcite is found in large massive veins, it lends itself appropriately to use as an environmental stone, excellent for birthing facilities, artistic studios and offices. It is also a powerful reminder of the right use of will for those who represent others as agents, managers and lawyers or who control others such as prison wardens and guards, school officials, judges, etc. *(2nd chakra)*

Peach or Salmon - This special variety of calcite is the higher octave of orange, and acts as an optimizer to open the lower chakras to energy from the upper chakras. However, it carries the hint of orange and willpower with it. This shade of calcite carries a tantric frequency, and empowers ecstatic and sacred dance. *(8th through 12th chakras)*

Pink - Pink calcite can be found in translucent, clear rhombohedrons with rainbows and perfectly cleaved edges. This mineral is the embodiment of etiquette and graceful manners. While orange calcite bridges seen and unseen as the ombudstone, pink calcite is the diplomat who makes way for detente, peace, prosperity and good relations. Pink calcite embodies love as a divine force, and transmits the courage to move through despair regarding global crisis to motivation to embrace hope and to dedicate to be part of the solution. *(4th chakra)*

White - White calcite improves memory, and augments the ability to astral travel when the awareness of how to do this with protection has been attained. Thought energy is amplified to specific body areas by this vibration of calcite, so it is an excellent mineral to support healing through visualization and biofeedback. It is a mineral which enhances the body/mind connection. White or clear calcite, even with a yellow tinge, is listed in the Etheric Cleanser division of the Spiritual Apothecary. *(All chakras, primarily 7th)*

CARLETONITE: Carletonite is a hydrated carbonate of potassium, sodium, calcium and silica. It is a rare mineral, but nice, small crystals are found in Mt. St. Hilaire, Quebec, Canada. It appears in fresh blue-colored rectangles with a white outline. It is an excellent sedative and relaxant. It is especially helpful for PMS and all water retention issues associated with hormonal shifts in women. It repairs the liver, and helps to counteract acidic conditions which exacerbate rheumatism and arthritis. It is especially effective in reducing bone spurs associated with calcium deposition due to arthritis. This is an excellent combination for healthy skin, hair, and fingernails. *(6th to 7th chakras)*

CARNELIAN: Carnelian is colored orange to brown chalcedony quartz found worldwide. Its orange to red/brown color range is caused by minute particles of hematite or limonite. Carnelian embodies the strength and fertility of the earth and the essence of jaguar madicine. It helps actualize personal power on the physical plane through second chakra activation. It grounds and focuses the wearer's attention to the eternal present. It helps to improve sense of touch, inner receptivity, and asthma, because it supports breathing deep into the diaphragm. It can induce deep feelings of comfort and serenity. It detoxifies the body from lungs through colon (fourth chakra down through first chakra). Complemented with visualization, it aids spinal cord and tissue regeneration. Though the second chakra is feminine, carnelian, due to its quartz/silicate structure, is masculine. Therefore, it balances second chakra functions, activates the adrenals (also helping with asthma), and energizes. It is useful to people who tend to poor circulation or feel cold when the temperature is sufficiently warm. It is an appropriate stone to provide energy to the aged, ill and weak, and aids right use of will. Carnelian, as a form of chalcedony, harmonizes well with all other chalcedony stones in a pouch or healing layout. Like other forms of chalcedony, carnelian repels evil (apotropaic). *(2nd chakra)*

CASSITERITE: Commonly called tinstone, cassiterite is a simple oxide of tin. It acts as a pollution neutralizer. The blackness of cassiterite allows it to strip pollution from the body without harming the body, and its reflective surface deflects toxins. It is a mineral which promotes "body talk," the communication by the body to the mind of the status of being, therefore it is useful to practitioners as an aid to diagnosis. It shares complementary uses with hematite. Cassiterite, applied to working with machines, provides complementary benefits. It helps machines to run cleaner, longer, and enhances the ability to troubleshoot mechanical problems. Cassiterite is a Healer in the Spiritual Apothecary. *(1st chakra)*

CAVANSITE: Cavansite is a hydrated calcium vanadium silicate. It is easily recognizable, once seen, as a bright and rich turquoise, needle-filled cluster. Pea-sized specimens from India look like little burrs. Micro specimens of cavansite have been found in Oregon. This mineral derives its essence from starlight. It expresses the crispness of starlight through boundary awareness and definition as well as the distance to reach the stars through detachment and the freedom from worldly concerns. It appears ever-new yet its pulsations are very old, providing an awareness of one's history yet a detachment and separation from its effect. It is always interesting and seldom unnoticed. It commands and arrests attention simultaneously, allowing it to enhance contemplation on the divine, the origin of life, one's purpose, and the ineffable wonder of the universe. For those with low self-esteem or a lack of confidence, cavansite is a welcome friend and antidote. It supports our relations to higher self and a more loving appraisal of self. Cavansite is a feminine and mother principle expression of vanadinite frequencies, therefore it supplies a type of discipline which is characteristic of how mother's raise children, and is an excellent preparatory stone prior to using vanadinite for a specific goal. It supports strength of character and feeling comfortable with self. It encourages safety in the company of others and maintaining one's boundaries. *(6th and upper chakras)*

CELADONITE: "Corpus stone." Celadonite is a lime green form of mica, and excellent celadonite in quartz specimens from Madagascar provide properties similarly as chlorite-included quartz. Celadonite and quartz is a prime all-body healer which when used with other stones increases their effectiveness or "amperage" significantly. Celadonite specifically addresses cell membrane permeability and helps to transmit oxygen to body cells. It provides transference capabilities across the physical/electromagnetic interface, especially through the bloodstream. Celadonite in quartz aids poor circulation, phlebitis and varicose veins. Like another mica mineral, muscovite, the molecular structure of celadonite has a vacancy which places it in the unique position of being able to get out of the way of itself as a healer. It helps to identify what's in the way so it can be addressed, and it strengthens the ability to remove oneself from the healing process so that pure energy can do the job. *(4th and 7th chakras)*

CELESTITE: Celestite, called celestine mineralogically, is a strontium sulfate categorized in the barite group. All forms of celestite are located in the Spiritual Apothecary in the Auric Sealers division. This mineral frequency wards off unwanted energies, frequencies, thoughts, entities and encounters. It shares with minerals in its division of the Spiritual Apothecary profound abilities to transmit meaning and pattern. Celestite, especially, enhances the ability to transmit significant information both accurately and charismatically. It is the stone of the "Great Servant," the martyr, the divine child, the sacred sacrifice and the holy, primordial sound "aum." From the Vakya Padiya, "All this universe is but the result of sound." All forms of celestite aid with psychic connection, especially precognition through dreams and clairaudience, as well as ecstatic witness of truth through prophecy and sermons. All forms of celestite are listed as Auric Sealers in the Spiritual Apothecary.

Blue (Madagascar) - Deep shades of celestite tune consciousness to higher planes, improving mental awareness of spirit beings and communication with them either during sleep or meditation, and provide auric protection during these times. Deep shades assist with transfer of soul in and out of the body during sleep, and aid deep breathing during sleep, thereby helping to counteract sleep apnea. Deep shades reduce the amount of sleep necessary to replenish and rejuvenate the body. Deep shades of celestite may reduce dream recall as rapid eye movement sleep is deepened, however this tends to result in increased abilities to actualize life purpose. They are recommended for areas where rest and meditation occurs, not for offices or kitchens. This form of celestite bears the vibration of spiritual truth and knowledge of one's true self. *(5th toward 7th chakras)*

Light Blue (United States) - This mineral is unique as a fifth and seventh chakra integrator, allowing abstract ideas to tumble from the mouth without hesitation. Celestite (strontium sulfate) is related to barite (barium sulfate) and anglesite (lead sulfate). Anglesite is considered the truth telling stone in action and at death, and golden barite is a mineral of truth telling in the written form. Celestite is the stone of truth in speech and hearing, and those who attempt to lie or obfuscate while holding it quickly become confused. The effort required to hold simultaneous realities often results in skin eruptions which tell the truth held within and spoken as a lie. *(5th and 7th chakras)*

White (United States) - White or clear celestite is a gem to behold. It sings so powerfully at an angelic level that it is highly prized on the other side as a literal, physical bridge between energy and matter. It is truth in thoughtform, and holds in its matrix the responsibility of how that thoughtform creates reality. This form of celestite's time is yet to dawn, but when it does, it will literally sing for all to hear. It is a powerful shamanic cleanser (exorcism) and protector. *(7th chakra and above)*

CERULEITE: These light blue nodules are a copper mineral combined with aluminum and arsenic found only in Chile. It works harmonically from the throat to the crown chakra. Ceruleite is excellent in healing aneurisms. It is also helpful as a contraceptive as it reduces fertility by bringing on anestrus, the pause between fertile times. *(6th chakra)*

CERUSSITE: Cerussite is a lead carbonate categorized mineralogically in the aragonite group. It acts on the higher self interface with the body. Many people are not aware of this higher self, because the connecting rods become clogged with negativities such as doubt and inadequacy and pressure to behave according to the status quo. These lines are cleansed by cerussite where the lead actually draws off unwanted energy. Because cerussite is lead-based, it is generally required that a crown chakra opener be employed with it in order to enhance the perception of connective links. Cerussite is listed in the Auric Cleanser division of the Spiritual Apothecary. *(Upper chakras)*

CERUSSITE and MALACHITE: The cerussite-malachite combination acts very much like tungsten. Of course, it is not metal. All electrical properties responding to the green ray in lifeform will benefit from interaction with this combination. It is a loosely combined aggregate which allows a great deal of air interaction to occur, so breathing with the mineral can cleanse and heal bronchial tissue. *(4th and upper chakras)*

CHABAZITE: Chabazite is a zeolite which forms a series with itself in three types. It is a hydrated aluminum silicate which can be either calcium, potassium or sodium rich, retaining all three of these elements in its structure regardless of elemental dominance. Chabazite has a blocky crystal structure. It enhances the absorption of water by tissues (good to counteract dehydration). It is excellent to balance all body fluid issues. It assists all health issues affecting joints, bones and muscles. It helps keep both the liver and kidneys in top condition. It aids in the integration of the two lobes of the brain. Visualize the phenomenon of frost heaving, how water crystals lift clumps of soil and create fantastic patterns, and you can see why chabazite is strong and magical in many ways, like frost heaving. *(7th chakra)*

CHALCANTHITE (Blue Vitriol): Chalcanthite is a hydrated copper sulfate, found most abundantly in Chile. Its vibrant sky blue color is unequaled as a natural or recrystallized growth. It is simply the stone of the Paraclete, Holy Spirit, Espiritu Sancti. It is the inspiring breath of life. The reason it is capable of such a high purpose is that copper provides the feminine electromagnetic pulse of earth life force energy while sulfur provides the masculine electromagnetic pulse of universal life force energy. The term vitriol spells out the initials of the alchemical formula which can be translated in two ways. Kurt Seligman, in his 1948 book *The Mirror of Magic*, translates *visita interiorem terrae rectificando invenies occultum lapidem* as "Explore the inner things of Earth and by distillation you will find the hidden stone." The hidden stone sought after in alchemy is the Philosopher's Stone. The initials can be translated from *visita interiorem terrae rectificando invenies operae lapidem* to: "Go down into the bowels of the Earth, by distillation you will find the stone for the Work." The work refers to the process of going deep within the self to discover that which is most enduring and to rebuild one's character and personality from this essence. Two other minerals carry the vitriol term from old time usage as well. Like chalcanthite, they are also sulfates, but melanterite, or green vitriol, is an iron sulfate, and goslarite, or white vitriol, is a zinc sulfate. Sulfur is an important element in alchemy as is mercury. See these entries as well as halite (salt) for more references to alchemy. *(3rd and 6th chakras)*

CHALCEDONY: Mineralogically, chalcedony is the name for a group of quartz varieties with compact, microcrystalline silicate structure. The types of rocks found under the chalcedony heading which are described in this book under their own names are: agate, jasper, onyx, carnelian, chrysoprase, bloodstone or heliotrope, and flint or chert. Chalcedonies listed below could equally be listed under jasper or agate, however they are usually classified commercially as a form of chalcedony. Chalcedony cannot be praised highly enough as a family of stones of great mystical power. Chalcedony is a Healer in the Spiritual Apothecary.

Black - "The Void stone." The entire chalcedony family is apotropaic, meaning designed to avert evil, and among the chalcedonies, black chalcedony is the most highly regarded spirit plane interceptor and guardian. It is unparalleled as a protection stone. Black chalcedony embodies the energies of the Hopi Kachina Crow Mother. It brings in the power of Mother Protector. As an emblem of the Dark Goddess or Black Madonna, black chalcedony brings in the energies of wisdom, fertility and the Void. It is a stone which enhances discipline, knowledge and discretion in the use of power, therefore it is excellent for parenting, mentoring and teaching. It is a stone of alchemical knowledge, the "good ground" which receives the seed of life and processes intuition from the bowels, the ground of being. It was common for spiritual leaders to revere this stone, therefore it is a member of the Spiritual Apothecary in the Auric and Etheric Sealers divisions. *(1st and 8th chakras)*

Blue/Purple - Blue chalcedony often resembles a periwinkle tone and fades to lavender. The distinction about when these shades differ is highly subjective. In blue and purple, chalcedony showers the head region with strong fourth and fifth dimensional properties enhancing a broad range of interconnections. This shade of chalcedony is a threshold or passage stone between worlds, realities, states and qualities. It mediates between elements, and can be trusted for its apotropaic qualities. It is a gateway marker between the known and unknown, the profane and the sacred. Blue/purple chalcedony is an important prayer stone and is most significantly used by the individual as an enhancer of spiritual communication, communion and integrity. It stimulates speaking in tongues, trance mediumship, and automatic writing. It is highly appropriate to give at times of spiritual commitment such as first communion, confirmation, marriage, and ordination. It is a stone of great thanksgiving and blessing, and is valued as a stone of extreme unction and ceremonial use at death. It supports letting go of the past. It embodies ineffability. This shade of chalcedony embodies buffalo medicine. Compare with Holly Blue Agate and Royalite. *(5th and 6th chakras)*

Libyan Blue - "Fairy's Breath Stone." Libyan Blue is found in a specific locality of Africa. By placing at tip of nose and rubbing the place at the back of the skull where the bones meet above the neck, one can stimulate the pituitary gland, and relax mental tension. For susceptible individuals, this same method can produce anesthetic effects. Good for sinus headaches and allergies, etc. *(5th and 6th chakras)*

Limbcast (Pink) - Limbcast is pink chalcedony which formed inside a tree limb cast, and the outside exhibits fossil markings of the limb. It is an energizer tuned to the subtle body or chakra system. In meditation, the stone promotes physical rest and auric stimulation. It is excellent for creative, meditative work. *(4th and auric chakras)*

Pink (Chalcedony Rose or Whetstone) - These stones are good to place in dishes of water or fountains to cleanse the environment as well as provide an indefinable pleasant aroma. The water should be replaced weekly. If placed in a sickroom, the water can be changed as frequently as twice a day (cancer) to every two days (minor operation). This mineral is especially attuned to the female physical body, and for healing disturbances regarding one's sexuality, if it is creating conflict in the individual. It is a pleasant meditation-companion stone which will activate a variety of subtle forces. *(4th and auric chakras)*

Sagebrush (Petrified): Petrified sagebrush is a peach colored form of agate called chalcedony which is thought to be a replacement of sagebrush. Called the "Smudger's stone" metaphysically, petrified sagebrush is an appropriate stone for dispelling confusion and negativity. It is apotropaic like all other forms of chalcedony. In conjunction with fluorite, petrified sagebrush dissolves released issues, ties, disease states and blockages. Since it travels well, it is excellent for clearing strange places prior to eating, sleeping or meditation. *(1st through 4th chakras)*

CHALCOCITE: Chalcocite is a simple copper sulfide with a gray metallic luster. It is heavy for its elemental composition. It is an excellent mineral for transformation consciousness, especially with respect to changing one's career mode from pinstripe to eclectic and non-traditional. It aids the desire to commit to humane actions and purposes. It resonates to Cosmic Christ consciousness. See Chalcanthite and Covellite. *(3rd and 6th chakras)*

CHALCOPYRITE: Chalcopyrite, called bornite mineralogically, is copper iron sulfide which turns rainbow colored during a mineral leaching process leaving a skin of multi-colored, powdery shades of iridescence on the outside. Keep specimens out of water and reduce handling as the outside "rust" will come off. It is called "stamina rock" by mineral spirit guides, because it is generally the best mineral to support long sessions of energetic healing, psychic work, meditation and creative output. It is an energizer, as its properties harness spirit energy to the physical plane. Chalcopyrite blends masculine and feminine energies of divinity and grounds them to the physical plane, so it naturally supplies balanced, sustained and continuous flow of energy. It can be trusted to provide energies of good judgment as it connects the root, heart and crown chakras. Chalcopyrite provides stimulus to the nervous system, brain, muscles, heart and lungs. It works simultaneously with the blood, bones and bowels. It is excellent for athletes engaged in competitive sports. It is a visual and physical stimulant, and is excellent as a preparatory stone for any day's activity involving heavy muscular and visual activities. This stone balances the superconscious, conscious and unconscious states of awareness, and will bring an overtaxed, hyperactive system to relaxation. Chalcopyrite can be used for closure ritual or releasing energetic ties after an intense day as it smoothes and grounds energy flow in the body, and aids the release of excessive amounts of energy. Chalcopyrite embodies the essence of elk. Chalcopyrite is an Auric Conditioner in the Spiritual Apothecary. *(1st through 7th chakras)*

CHAROITE: Charoite is a hydrated potassium calcium sodium silicate with fluorine. This massive, purple and black mixture is found in Russia. It is a stone which emanates classical vibrations akin to Bach, Beethoven or Chopin. Though it has a relatively simple chemical formula, it bears the structural ability to integrate complex waves of energy into a masterpiece of accessibility and purity. The sensibilities of this mineral are finely tuned, allowing it to support fine nervous system activities and to purify conditions, whether physical or environmental. It is highly adapted to support creative work. It stimulates the ability to integrate rapidly changing energies, which can be the result of shocking circumstances on earth or effects from off-planet energies such as astrological cycles and discoveries in astronomy. Some people will respond strongly to this mineral, signifying resonance to the mineral's integration characteristics. Charoite contains power in subtlety, the majesty of being able to evoke increasing levels of awareness based on being able to integrate increasing complexity of insight. While it encourages depth of surrender and search, it opens to larger fields of beauty and truth. Charoite speaks naturally of bridging what is viewed as wild and unattainable, even threatening in its prospect, with what is most real, most whole and most visceral within the self. Therefore, it occupies a position of being on the edge where diverse realities meet, share and integrate to create new species of thought, artistic expression, and eras of human endeavor. Charoite helps to regulate body fluid balance, supports liver, heart and kidney function, and aids the elasticity and tone of muscles. It regulates irritability and excitability of muscles and heart. It reduces acidic conditions, and helps arthritis and gout. Charoite is a Healer in the Spiritual Apothecary. *(6th chakra)*

CHERT, MULTI-COLOR: Chert is a member of the group of rocks classified under the heading of chalcedony quartz. It is flakier than flint, but chert can be regarded as the same type of rock as flint within a spectrum of hardness. It comes in a variety of colors. This stone is good for improving humor. It has a calming effect. It redirects energy. It protects. Like other chalcedonies, chert absolutely stops negative vibrations, therefore it is excellent to utilize in preparing for and during healing work. It works in harmony with all other stones. When kept near the head at night, it brings light-filled and peaceful dreams. It is beneficially inclined for the hyperactive person. Very young children will sleep easier when this mineral is near them. It balances and aligns the five bodies. It will help the holder to release unwanted and unneeded burdens or blocks. It aids in unblocking clogged energy points. One of its companion plants is rosemary. It is called "sedovite" on the spirit plane, as it enhances serenity. Best results occur in bigger sizes (five to six inches and larger). Chert is located in the Etheric and Auric Openers division of the Spiritual Apothecary, as it provides protection as well as alignment and balance. *(All chakras based on energetic properties, not color)*

CHIASTOLITE (Cross stone variety of Andalusite): Andalusite is an aluminum silicate, and chiastolite is a metamorphic variety characterized by cruciform inclusions of carbonaceous material. It is a good protection mineral. Chiastolite strengthens balance and spiritual discipline. It is a good centering stone, and is especially powerful when used in a mineral totem pouch. *(1st chakra)*

CHILDRENITE: Childrenite is an iron aluminum phosphate, and forms in a series with eosphorite. It is fairly rare, and comes from Brazil. It crystallizes in the brown to orange/yellow range. The name of this mineral is an appropriate match with its qualities. Its short, flat needles of dark golden appearance suggest third chakra applications, such as enhancement of self-knowledge and release of past life burdens and karma which may overpower higher purpose in this life. It allows a good, fresh start, and the conviction to carry out dreams and goals, ideals which are wished for children all over the world. It provides blood strength in the form of courage, valor, loyalty and bravery. Its qualities are enhanced by the presence of copper minerals, such as malachite, azurite and dioptase. *(1st through 3rd chakras)*

CHROMITE and MALACHITE: Chromite is iron chromium oxide classified in the spinel group. By itself, chromite is a singer variety of mineral which energizes clairaudient capacities. It is capable of enhancing frequency perception. It is also a lung healer much like malachite by itself. Chromite tends to induce coughing, and is good for expectorate needs. It should be followed up with obsidian and turquoise placed on the chest which stops the coughing reflex. *(4th chakra)*

CHRYSOBERYL: Chrysoberyl is a beryllium aluminum oxide. Chrysoberyl can be colorless, green, yellow, gray or brown. Alexandrite is the gem variety of chrysoberyl and flashes red in artificial light and green in natural light. A cat's eye variety of chrysoberyl contains microscopic inclusions of rutile. As a gemstone or mineral worn on the body, chrysoberyl aids kidney and liver function by removing toxins and accumulated debris, therefore it is excellent for dialysis patients and for anyone suffering from liver disease, kidney stones, and kidney disease. It has sympathetic effects on the pineal gland, releasing toxins and toning gland tissue. It is a useful choice to aid clearing the body after psychic work, especially for those people who do pollution and fault pressure release work geobiologically or geopathologically. It is an excellent diagnostic tool, and aids individuals to "know where to go" to do healing work with the earth or with individuals. Chrysoberyl is a Healer in the Spiritual Apothecary. See Alexandrite. *(3rd chakra)*

CHRYSOCOLLA: Chrysocolla is a hydrous copper aluminum silicate. It carries a blue to bluish-green to turquoise color range. When quartz-rich, it takes a good polish and is termed gem silica. This is a stress release stone, and an emotional and mental balancer. It strengthens the lungs, and the pituitary and thyroid glands, and it reduces anger and fear. Because it melds the abilities of the heart to both learn and implement with the abilities of the brain to learn and understand, chrysocolla maintains a "level-headed, open-hearted" stance which few other minerals can provide. It enhances biofeedback accuracy and aids in conscious control of autonomic functions of the sympathetic and parasympathetic nervous systems. Chrysocolla stimulates feelings of deep safety and trust. It enlarges the ability to charge one's thought process with positive energy in the form of praise, gratitude and serenity. It is an excellent preparatory stone for healing work. As specimens of chrysocolla increase in size, other minerals and elements tend to mix in and alter the energetic dynamics, therefore smaller chunks, usually prepared for jewelry, retain pure chrysocolla characteristics. Chrysocolla is found in the Spiritual Apothecary in the Etheric Opener division like its copper cousins azurite and azurite/malachite. See Gem Silica. *(4th and 5th chakras)*

CHRYSOLITE (Gem Olivine or Forsterite): Almost all olivine is actually forsterite, which is a magnesium silicate. Chrysolite is an old timers name for olivine or forsterite. It is a green to golden stone of the sun which stimulates creativity and spiritual insight. It is best worn at the throat chakra. It reduces depression and conditions of the body which are generally toxic. *(5th chakra)*

CHRYSOPRASE: Chrysoprase is a bright green to yellow variety of chalcedony quartz containing traces of nickel. It comes primarily from Australia, and is often found with brown, iron-rich rind.

Dendritic - This aspect of chrysoprase is excellent in balancing carbohydrate metabolism in humans. It is especially useful for diabetics, but hypoglycemics will find it beneficial as well. The black branches or dendrites are manganese oxide. *(3rd chakra)*

Green - Better grades of this quality are revered as if they were a fine, imperial jade. It enlivens insight for resolution of personal problems, and increases selflessness and love in actions. Its crisp-looking green presents a high octave for its energetic range. Like other chalcedonies, chrysoprase is apotropaic, and possesses the energetic clarity to release stress, negativity, unwanted energies or entities and attached thoughtforms. It facilitates changing unwanted behavior patterns through increased self-awareness. It tends to attract the benefits of positive thought, and brings spirit response for meritorious behavior and conscientious discipline to activate spiritual principles in one's life. It is the stone of the peacemaker. For some body frequencies, it is more applicable to healing work than chlorite-included quartz, but this is not overly common. Chrysoprase is found in the Spiritual Apothecary in the Auric Cleanser division, and can be substituted for chlorite-included quartz in the Etheric Conditioner division for those who respond appropriately. *(4th and 5th chakras)*

Lemon - A mineral which provides maximum utility in rainy, foggy or dark environments such as at the poles or in extreme northern or southern latitudes. Good for depressed or isolated individuals. The mineral attracts positive vibrations to its vicinity. Both the yellow and lime green varieties are called lemon chrysoprase. *(3rd chakra)*

CHRYSOTILE: Chrysotile is a hydrous magnesium silicate in the kaolinite-serpentine group. It is more appropriately known as clinochrysotile or orthochrysotile. It is a fibrous variety of serpentine, more commonly called asbestos, which forms in a spiral. This mineral can be yellow, green, gray, white or brown. It is a type of serpentine which is typically used to draw out body toxins. The yellow color variation would tend to work best on the abdominal and lower areas. It is an ingredient of the cat's eye family of minerals. See Serpentine. *(Usually 3rd, 2nd to 4th chakras)*

CINNABAR: Cinnabar is a mercury sulfide. Cinnabar links East and West traditions of alchemy, because it is made of the two elements important to each tradition. From the tradition of Western alchemy, we know that mercury represents the soul and sulfur represents the spirit, therefore cinnabar bridges what is honestly real and deep within each of us to what is supremely divine. In the East, cinnabar and mercury represented rebirth and the ability to regenerate. Together they symbolize the Phoenix rising from the ashes. Eastern alchemy ignored sulfur as a spiritual element, and focused on extracting mercury to obtain the substance they believed conferred immortality or at the very least long life. However, just the opposite was true. It was not uncommon to ingest mercury which led to early death and brain dysfunction. Today, cinnabar is crushed and reconstituted for jewelry purposes, but some natural specimens are available. Cinnabar is good for conditioning reflexes as it provides feedback to the autonomic nervous system regarding physical exercise and practice. It is useful for all professions or interests which involve developing reflexes, such as race car driving, hunting, martial arts, etc. With clear quartz, cinnabar can be utilized to strengthen the nervous system and brain wave polarities. Cinnabar reduces and balances overactive electromagnetic impulses, and can be helpful for sleep apnea, Turrette's Syndrome, and Parkinson's disease, among others. Cinnabar was used historically as vermilion pigment. Cinnabar is listed as an Etheric Conditioner in the Spiritual Apothecary. See Mercury and Sulfur. *(1st and 12th chakras)*

CITRINE: See Quartz.

CLINOCHLORE: Clinochlore is in the chlorite group, and it is a magnesium iron aluminosilicate hydroxide. It is an essential component of many chlorite and talc schists. Clinochlore crystals are rare, blue-green, and shaped like a barrel. Massive clinochlore is called penninite and looks like mica, because it lays down in a lamellar fashion. It can be whitish to green. Both types have the same chemical formula. Clinochlore gets to the heart of the matter electrically. At the moment of fertilization when egg is penetrated by sperm, there is an electric pulse which transmits throughout the body of the woman. This pulse of life occurs in many life processes and is usually unnoticed. When the moment requires intense presence or intention, clinochlore delivers strength, attention and foundation whatever the occasion. It facilitates electrical connection from the nervous system to the blood, muscles and bones while maintaining grounding. It aids the endocrine system, enzyme function, and waste elimination. *(All chakras)*

CLINOCLASE: Clinoclase is a copper arsenate hydroxide, and it contains one of three elements which can be classed as serpent-based. The arsenic in clinoclase, being toxic, places this mineral in a category of healing which derives from metabolizing what can be classed as venom in the mineral kingdom. Do not rub this mineral on the skin, inhale or lick it. Utilize its properties energetically. Clinoclase relates to serpent power and symbology. It is a healing mineral for the spinal column and facilitates movement of kundalini energy. It is a cold-blooded mineral, meaning it often provides maximum benefit after it has been warmed in the hands, briefly over candle flame or gently in the sun. *(1st to 6th chakras)*

CLINOZOISITE: Usually colored a drab green/brown or pale green or pink, clinozoisite is a calcium aluminum silicate hydroxide often found associated with garnet, in particular grossular garnet. It is dimorphic with zoisite, meaning they share the same chemical formula but they grow in different crystalline structures. They are both members of the epidote group. Clinozoisite is excellent for detoxifying the body of radiation poisoning. It can be worn or used as an auric comb, but should be cleansed in or exposed to sunlight regularly. Like thulite, clinozoisite offers properties of emotional bonding, surrender to safety in intimacy, nurturance and emotional loyalty. Like epidote, clinozoisite offers emotional clarity during times of stress and crisis. It is applicable to mending heartache, especially in the young. Clinozoisite is an Auric Cleanser in the Spiritual Apothecary. *(4th chakra and aura)*

COAL (Anthracite, Bituminous, Lignite): From a softened brown to a hard black glassy anthracite, the stages of carbonized organic matter form a broad density range in this mineral class. Following the principle of like aiding like, it can be used as a remedy for toxic conditions caused by petroleum products and radiation contacts. Body tissues are actually stimulated to cause regeneration, allowing for an uplifted feeling as healing occurs. *(1st chakra)*

COLEMANITE: This mineral is a hydrated calcium borate which crystallizes and is semi-hard. For those who cannot integrate with danburite immediately due to danburite's extreme shamanic power, colemanite is an excellent mineral to gauge oneself slowly and deliberately into the shamanic realm. Colemanite is more fluid and forgiving than danburite to an initiate's first work. It is no accident that the premier flight test location in the world is located adjacent to the largest deposits of boron and borax discovered globally. Boron is an important component of rocket fuel. The environment of the Mojave Desert is suffused with tremendous light-carrying capacity and the element boron, which strengthens human nervous systems to withstand high levels of stress. Colemanite provides unexpected benefits during times of need. It can be asked for help, but generally the frequency attunement is to subtle body levels beyond conscious awareness. Colemanite serves the higher self and allows interaction to flow from spirit to physical. See "Lagniappe" in dictionary. *(7th and subtle body)*

COLORADO SUNSET: "Adventurer's stone." This is a newly found orange quartz-like rock with hematite bands. This mineral combination has a strong association to leadership, empowerment, risk-taking, and initiative. It is naturally grounded, providing a base of confidence and security to new activities or endeavors. If the energy of the stone appears harsh or unbalanced to the wearer, it can be worn with stones which promote tolerance, patience, and compassion. *(1st and 2nd chakras)*

CONICHALCITE: Conichalcite is a calcium copper arsenate hydroxide. The green mossy skin typically found covering this stone provides a mental and energetic link to the plant kingdom. Patients on chlorella therapy or spirulina users will strengthen the effect if conichalicite is worn while using these algae substances. Conichalcite is an overall healer mineral because of its copper basis. Copper affects all the chakras. Its skin-like signature suggests its most powerful specific use: healing skin problems such as psoriasis, cancer, herpes, and fever blisters. Its arsenic property provides a serpent healing energy conduit which specifically helps to combat "unhealable" diseases and viruses. *(Skin, 4th and all chakras)*

COOKITE: Cookite is a lithium aluminum silicate in the chlorite group. It forms in aggregate stacks. Cookite is often found intergrown with clear quartz crystals. It is an excellent balancing agent for a form of manic depression called cyclothymia. This mineral enhances resonance by clarifying and sharpening output of electromagnetic signals in the body. *(6th and 7th chakras)*

COPPER: Copper is a metallic element, and was the first metal to bring humans out of the Stone Age, because it was so easy to extract, process and work. Near nickel and above both silver and gold in the Periodic Table of the Elements, copper and the others are known as the "coinage metals." Copper is named after the Latin, cuprum, or Cyprus where copper was mined. Copper is associated astrologically with the planet

Venus, however it is more accurate today to acknowledge that copper is truly an earth energy conductor. It symbolizes plant life, and in many cultures, red and green both represent the life force.

Copper balances the subtle and physical bodies, as well as body electromagnetics. It may be essential to bring life force energy to DNA in genetic engineering (*The Cosmic Code* by Zechariah Sitchin). One of the two most readily utilized metals in the body, copper is necessary for bones, but it is essential to properly utilize iron, the other important body metal. A small amount of copper catalyzes the process of iron assimilation, so it is helpful for all ailments where iron is lacking. Copper improves metabolism, reduces inflammation, and increases blood flow. It helps with arthritis and rheumatism, and is a valued poultice supporter for infected wounds. Save your stray pennies in a copper or brass bowl by the phone to set up a resonant field of clarity in communication, balance and calm. If telephone encounters tend to be taxing or stressful, copper by the phone increases intention to maintain boundaries with telemarketers and friends who require a lot of advice. Several cultures believe that copper reveals the mundane meaning of conversations and words while gold reveals the deeper, secret words of power. Copper is believed by some indigenous people to be implanted by bolts of lightning which are in turn placed in the ground through thunder.

Copper is one metal which symbolizes justice. Both the Russians and Haida of British Columbia believe that the Mistress of the Copper Mountain or Copper Woman, respectively, have green eyes, one of two colors (also blue) that copper creates in minerals. Copper Woman, according to the Haida, created humans from her body fluids after she came here from outer space (Venus?), lending cosmological credence to Sitchin's suspicion that copper is necessary for successful genetic engineering. Copper is located in the Auric Conditioners category of the Spiritual Apothecary as it infuses the subtle and physical bodies with light and energy. *(All chakras)*

COPROLITE: Found in Washington, coprolite looks like what it is (was): fossilized dinosaur dung (a variety of amphibious turtle). It is largely metallic and rusty looking. Believe it or not, coprolite is good for stimulating bowel movement. It promotes water retention in the bowel, and release. It has an effect opposite to its state. On psychological and subconscious levels, it supports relaxation and overcoming attachment issues. *(1st chakra)*

CORAL: Calcareous, all corals have complimentary attributes. All forms of coral help fever, colds, asthma, and digestive problems. They are especially attuned to the lungs and digestive system. Because coral grows like a tree in water, it is a special symbol of the Tree of Life emerging from the origin of life. It is unique as a growth which functions in all three kingdoms: animal, vegetable and mineral. Corals are excellent for deep thinking and deep thinkers. The State stone of Florida is agatized coral. All forms of coral are listed as Auric Sealers in the Spiritual Apothecary, because they exhibit the ability to thrive in a hostile environment.

Black - Black coral is protective psychically and enhances abilities to sense psychic information on a physical level. It is useful for psychometry, as well as long distance viewing. It helps to settle nervous upset. It can be used as a "spirit caller," as black coral transmits spirit support to humans. It possesses a feminine vibration. Black coral is the state gem of Hawaii. *(1st chakra)*

Blue - Blue coral calms and attunes life forms to Earth energies. Blue is a signature color of sentient life, and signals to other life forms that assimilation is occurring. Therefore, blue coral supports physical absorption of nutrients as well mental absorption of energy and ideas. It provides a masculine vibration. *(5th and 6th chakras)*

Pink - Pink coral speaks to the delicacy of moving forward in space and time either through the ability to create new life and birth it successfully or through the ability to receive spirit connection and offer what has been transmitted to others, such as through inventions, prophecy, creativity, etc. Pink speaks to a willingness to hold space which is not passivity nor is it assertive control. Pink is a coral of mediation, forgiveness, intent and trust. It emits a feminine vibration. *(4th chakra)*

Red - Red coral provides physical protection and energy, and acts as an astringent to mucus membranes. Red coral stimulates the root chakra, and vitalizes life force energy. Red and white coral together provide strength for creativity. It possesses a masculine vibration. *(1st chakra)*

Red Horn - Red horn coral is specifically attuned to physical science inquiries such as physics, chemistry and mathematics. The red tone draws thoughts through the tissues of the body to ground currents which trigger response pathways in humans, allowing conclusions to be supported or refuted. The thinker may be completely unaware of this process, but working with red horn coral can increase sensitivity to the whole body process of intelligence and knowing. *(6th and 7th chakras)*

White - White coral contains the spectrum of color and energetic possibilities, therefore it supports balance, clarity, diversity, and unity. It is inherently sacred bringing to the oceans a visual and physical receptacle for spirit to infuse. It is like a beacon of salvation, and signals the presence of divinity. White provides a neutral vibration. *(7th chakra and above)*

COVELLITE: Covellite is a copper sulfide. It is an iridescent copper mineral with faces of indigo to purple sheen. Some of the world's finest specimens came from Butte, Montana, but it is now extinct in its prime mining areas. Covellite is a stone of energy and intent like chalcopyrite, but with its more specific color or action, it works on the mental area, giving force to creativity and all mental pursuits. It is a good stone to use when meditating, working out complicated ideas or problems, or contemplating how to accomplish a goal. See Chalcanthite and Chalcocite. *(6th chakra)*

CRANDALLITE AND VARISCITE: Crandallite is a calcium aluminum phosphate and variscite is an aluminum phosphate. Crandallite forms the yellow portion only rarely found in some variscite nodules, which are mostly green. This combination aids the process of healing DNA abnormalities to prevent disease and transmission to the next generation. This combination is very potent for humans rather than animals. It soothes the nervous system, decreasing effects of stress and allowing deep relaxation. It is balanced energetically between yin and yang or masculine and feminine vibrations. *(3rd and 4th chakras)*

CREEDITE: Creedite is a hydrated calcium aluminum sulfate flouride hydroxide. The small crystals that form this mineral are usually found in light lavender hues embedded in a white matrix. The internal crystal lattice network is very condensed, enough to make a psychic mind squint, and therefore has energy compacted and stored for future use. Occasionally, it is found in a dark violet form which takes this power to above the third eye in an area called the "high mind" center. Here our intuitive powers of the sixth chakra are connected to spiritual opening in the seventh chakra. This will be an important area to activate in the "unfolding" times happening on Earth. *(6th through 7th chakras)*

CROCOITE (Eugenic Harmonizer): Crocoite is named for the color of a particular crocus flower. The single best locality of crocoite is Dundas, Tasmania, Australia, and it glows with a rich, orange-saffron color in pencil-shaped crystals. It is a soft lead chromate mineral which is quite heavy for its size. Crocoite is one of the best eugenic (the science of improving inborn qualities, genetics) harmonizers available to man. Used by women, it facilitates the knowledge of proper conception time, and allows the highest use of astrological planning. As a full-time mineral worn by a single female, it enhances her ability to evaluate the complementarity of the male as a mate and father. The full-bodied color is reminiscent of a healthy, fertile female, and promotes fecundity in women, thereby aiding women to fully engage and flow in feminine aspects. Crocoite is a feminine power mineral which supports release of sexual abuse trauma and the ownership of receptivity as a basis of life-giving power and integrity. Crocoite is listed in the Etheric Conditioners category of the Spiritual Apothecary, because it infuses the gene structure with life force energy. See also Calcite, Orange. *(2nd chakra)*

CROTIA "DIAMONDS": Crotia "diamonds" are found in small groups of pea-sized, gray quartz. They provide an excellent crystalline structure for bonding the light and dark aspects of self into a synergistic whole, like hanab ku, shri yantra, or yin/yang. Crotia "diamonds" incorporate the shadow side into the whole structure. *(1st and 7th chakras)*

CUBIC ZIRCONIA: This stone is man-made, but it has some energy activation powers. Its rainbow of colors affects the whole being and the crown. *(7th chakra)*

CUPRITE: Cuprite is a copper oxide normally found massive and mixed in with azurite, chrysocolla and malachite. A rare crystallized octahedral form has been found in South Africa. It bears a burnt or brick red color. The chemical formula of copper and oxygen is highly usable by the human body, and works well with the subtle bodies.

Crystalline - Cuprite's dark, ruby-colored crystals, along with its copper basis, help with vitamin and mineral absorption. Though a stimulating primal red, it works on the heart chakra and chest region in an opening, and then balancing, function. When cuprite is present in specimen form with other copper minerals, it can be an excellent bedside companion for critical care situations. *(4th chakra)*

Massive - This form of cuprite maximizes digestion and absorption of nutrients by affecting the base, second and third chakras. It provides a melding influence, and can be quite relaxing compared to the crystalline form. It can be effective in relaxing the individual enough to overcome colitis, acid reflux and gout on the physical level, as well as performance anxiety on the psychological level. *(1st, 2nd and 3rd chakras)*

CYANOTRICHITE: Cyanotrichite is a hydrated copper aluminum sulfate hydroxide. Eons and eons of moons ago, sages walked the planet as if young children danced in them, yet they were then very, very old. One day Hawk called down to Nambia, a wise panther of South America, and asked where all the mice had gone. "Why, Hawk," answered Nambia, "they have crossed over the great sea to Asia to better know their numbers." Hawk flew off to Asia and some time later returned. "Nambia," said Hawk, "the owls of Asia say they have eaten all the mice and wonder when you will send more." Nambia noticed Hawk looked thinner and worried. What shall I do with these ignorant creatures who know not how to preserve their numbers? asked Nambia to himself. A great wind came up and ruffled through Nambia's fur, whispering a soft message. "Go within the earth, Nambia. Find yourself a cave and dream. Let your breath be breathed by Earth that she might become your wisdom and in turn teach her children." Nambia had never been in a cave before, and couldn't figure out how Earth knows wisdom from breath, but Nambia trusted the wind, because wind spoke with force or gentleness when needed. Nambia searched many long months for the cave and couldn't help wondering who would live to tell the tale. What made Nambia stop and finally trust the cave in which she entered was the skeleton of a mouse she found at the opening in some rocks. As she climbed into the hole, she felt dizzy and weak, and immediately was swallowed up in a slumber from which she never woke. The legend has it that Nambia became essence which became cyanotrichite which taught Earth to know herself and her children to know survival. See Chalcanthite.

DANBURITE: Danburite is a calcium borosilicate. The finest specimens come from Charcas, San Luis Potosi, Mexico. In Spanish, danburite is called flor de hacete, flower of the hatchet, which aptly describes the shape of its termination. Danburite is a member of the boron family of minerals, all of which possess shamanic characteristics. It is the premier shamanic tool, and is generally best used in the hands of an experienced crystal practitioner. See Colemanite, Datolite and Ulexite for assistance in building endurance to utilize danburite. It can be programmed for highly specific precision uses such as tissue regeneration and tumor removal in psychic surgery. It can be placed with healing tools to strengthen their qualities and maintain their potency when being used frequently.

Danburite is placed in the Auric Openers division of the Spiritual Apothecary, because it allows huge amounts of light energy to be passed through the subtle and physical bodies. The drafts of light can be sudden like bolts of lightning, however boron provides energetic stability to the body so it can withstand intense light pressure as it transits the physical and subtle bodies. Lightning is an ancient symbol of intuitive and spiritual enlightenment, the spark of life, fertilization, and that which created life.

The Greek Zeus, the Old Testament God, the African Pygmy sky-god and the Australian Aboriginal god among many others all formed life through lightning, which was often depicted as seminal ejaculation. Boron is the first and important non-metal element which incorporates to form living organisms. Through the presence of boron, danburite is harmonically tuned to three sacred plants all of which contain boron: maize, saffron and grape. Ancient cultures recognized these plants as sacred for good reason. Danburite resonates to thunderbird energy. *(7th chakra)*

Blue/Gray - This Bolivian (Alto Chapara) form of floater danburite comes double terminated, and is usually an inch long. This tone of danburite is a stone of intense charm, grace and openness. It instills a sense of belonging, connection to others and awareness of one's presence in the whole. It facilitates precise tenderness, the knowing of how to comfort others and put them at ease in stressful situations. Blue/gray danburite is useful to those who work for decision makers or elected officials as it supports those who act as the power behind the throne, the trusted advisor, or the one who controls access to an official. See Aragonite (Cuprian). *(4th and upper chakras)*

Clear - Note that this mineral looks like a combination of tourmaline (the striations and borosilicate chemistry) and topaz (the termination), yet it is clear like quartz. This form of danburite was used at the turn of the 20th century as an afforable option to diamonds. In the natural state and as a faceted stone, rainbows flash out of its clarity energizing the subtle and physical bodies. Clear danburite is pure, unadulterated, "let's get moving" yang energy, capable of sending immense power and strength. Organized in mineral structure, lightning is ordered and directed in a highly specific pattern. Used unconsciously or cavalierly, it can bring down the destructive potency of lightning. A little danburite goes a long, long, long way.

Having said that, clear danburite's name, flower of the hatchet, expresses danburite's archetypal meaning of lightning at death and shamanic initiation. People who die by lightning strike are said to be immediately enlightened and sent to heaven. Returning to life on earth after death by lightning means a person reincarnates as a shamanic adept. This is the mineral of the spiritual warrior. The hatchet or lightning axe is included among the symbols of all sky gods. The diamond mace, or adamantine, represents transcendental truth, and the rolling of the Dharma thunder signifies liberation teaching. Blasting forth from the third eye of Shiva is the cosmic intelligence of the thunderbolt. The vajra and dorje, both symbols of the thunderbolt, are represented also by styllized flower shapes, or the fleur-de-lis, a conventionalized iris. It is not known who named danburite "hatchet flower," but the appropriateness of its rich symbolism is stunning. *(Primarily 7th chakra)*

Milky - This is the balanced yin/yang version of danburite, and though more common and less expensive, it is a complete shaman's tool. Milky danburite connects the self to internal light, helping the eyes to shut to outer distraction in order to meditate. It increases equilibrium and the ability to listen as well as visualize. It symbolizes the sacred union of sky and earth. It cleanses evil, and frees the soul to seek Light and Truth. *(All chakras)*

Pink - The female shaman's stone, pink danburite, can be used by either males or females to enhance yin powers. Using both clear and pink danburite promotes balance of yin/yang powers and energies. Pink danburite is also a shapeshifter's power tool. Pink danburite is safer to use when invoking fire and rain energy for shapeshifting, because it naturally brings in the protective, life-giving attributes of lightning. It is associated with the Virgin of Light. Pink danburite can be thought of as the bringer of life, as in the ancient symbol of stork. Stork was considered a powerful creature to Taoist alchemists along with raven and hare. Studying these symbols allows us to recognize the life-giving story about pink danburite being told in symbols. Stork is a symbol of spring, fertility and causer of pregnancy. Raven is the messenger of the gods of thunder, and raven's tongue is said to be a lightning bolt. Raven transmits long distance healing. Both birds are considered symbols of family strength, because they feed their old, and ravens work as a family unit to raise young. Hare is a lunar symbol, because it comes out at night, and it is unusually prolific. Hare's feet were considered lucky and a protection against black witchery. Hare is a symbol of rebirth, intuition, light in the darkness, dawn and new beginnings. Hare is a yin symbol as well as a symbol of self-sacrifice, because when Buddha was hungry, hare offered itself and jumped into the fire. All these attributes are found in pink danburite. *(4th through 7th chakras)*

DARWIN GLASS: Darwin glass, also called heat or lightning glass, is olive green and comes from Tasmania. A mineral of compelling character, Darwin glass is a fine specimen for doing the work of Moldavite tektite at less cost. The glass stone is the product of soil vitrification when a meteorite hits earth. It is as or more conducive to third eye work than moldavite. It is especially appealing as a mineral to enhance contact with dolphins in physical form. *(3rd, 4th and 6th chakras)*

DATOLITE: A calcium borosilicate hydroxide which forms in the same environment as danburite, datolite can also appear with zeolites, prehnite and calcites. It is a shamanic mineral. Datolite is the correct hardness among the shamanic minerals to induce physical warming and awareness of light charging the physical body. It is a good tool for a client to hold while a practitioner reads and diagnoses a client's subtle body fields, because it increases definition of subtle body layers. During some forms of energy work, clients may feel cold, so placing datolite near the person will increase warmth. Datolite is an intermediary between colemanite and danburite for the shamanic initiate. It is excellent for kicking old habits and patterns which are no longer useful or desirable to the individual. It supports the subconscious in receiving positive affirmations to establish new directions and lifestyles. It supports accessing the subconscious during any activity with it, such as hypnosis, meditation, creative work, subliminal programming and dream states. It is beneficial to use with programmed clear quartz. Its lime green color is a higher vibration of heart chakra green. *(All chakras)*

DATOLITE NODULE: Though usually lime green crystals from Connecticut, Mexico, and Russia, this Michigan form is always found in solid nodules mixed with copper (red part). The Michigan variety clusters in pillow-like crystals. Datolie nodule is excellent as a gastric stimulant and counterirritant. *(1st chakra)*

DESCLOIZITE: Descloizite is a lead zinc vanadate hydroxide. A great motto to describe it is: Out of harm's way and into the arms of play. This mineral is attuned to the chemistry and awarenesses of children and the child self, therefore it is particularly beneficial as a protection stone for children, and as a reminder of the spirit of play and freedom. Descloizite activates endorphins and supports the endocrine and immune systems. It is a powerful mineral for shielding the common cold and flu, and it is an excellent choice for activating intention. It is much more forgiving in the area of intention manifestation than vanadinite, and is a good place to begin practicing with the higher integrity demands of vanadinite. Descloizite is the divine child complement to vanadinite (father principle) and cavansite (mother principle). *(1st chakra)*

DIAMOND: Diamond is the element carbon in its highest expression of clarity, perfection and gem quality. Diamond provides protection from poisonous and toxic influences. It overcomes evil and fear, and enhances courage. No emotion can affect its pure mental nature as it reflects emotional states back to their origins. Diamond shields the crown chakra from mental and emotional excess. It resonates to spiritual unity, and represents spiritual invincibility. It is a premier conduit of spirituality as it expresses a noble and complete consciousness as a mineral. It is said to free the spirit from attachment to the phenomenal world, lust, and anger. It brings truth, wisdom, loyalty and moral integrity. In Eastern tradition, diamond is the symbol of Buddha, the Void, immortality, and enlightenment's razor edge. In Tibet, the word *diamond* means "the queen of stones." The diamond represents immutability, constancy, no fluctuation; therefore, Buddha sits on a diamond throne and the highest ceremony is the Diamond Sutra.

It is notable that diamond is pure carbon, the element which until recently was thought essential to living forms, until silicate-based lifeforms have been discovered in the deep oceans. Nonetheless, life is an emotional experience. Returning to the light is a detachment from emotion, a shedding of emotional reality, and a return to a state of complete light and clarity. Diamond expresses the yearning of living organisms to return to the light and to achieve the pinnacle of development. It is amazing the hardest crystal known to man originates from life itself, from the most fragile to the most rigid. It is a lesson as well that Christ counseled, "the meek shall inherit the earth." What is likely meant by that statement is that rigidity leads to the inability to adapt, and flexibility ensures survival. To be a mobile living form is not the same as being a crystalline mineral which transmits its character as a foundation stone for the benefit of those returning to the Light. Why are diamonds a girl's best friend? Because they provide emotional shielding to the feminine receptive nature, and increase the capacity to express emotions clearly and in a detached manner. Diamond is the state gem of Arkansas. Diamond is the April birthstone. *(7th chakra)*

DIAMONTINA QUARTZ CRYSTAL (Star Seed Crystal):

Clearer Diamontina - Is like the difference between blue and brown eyes. It is like blue eyes, and carries within it psychic serenity and celestial interaction. This quality of diamontina is meant to enhance the earth/cosmos connection. Cracked faces within a crystal are frequency receiver points. As they change, so the frequency changes, usually to become stronger and easier to detect with practice and familiarity. These crystals have been used before in past life, and are usually frequency specific — meaning the frequency is attuned to that of the individual user. Being used before can mean by that individual or anyone else. *(7th chakra and up)*

Clouded Diamontina - Is like brown eyes, doelike and well grounded. These crystals are meant to enhance physical body working on the planet (any planet or star). The crystal allows the auric and etheric bodies to provide immediate transformational input to the cellular structure of the physical body. This is programmed and activated by the subconscious/ higher self interface. Information is "downloaded" to the crystal where it can provide benefit. Clouded crystals will often clear with use, which denotes that the user is clearing toxins from the physical body. This process allows all aspects of the being to integrate more fully. As with clear diamontina, clouded have been used before, and are frequency specific. *(7th chakra and up)*

Double-Terminated Diamontina - Extremely powerful tools for all beings. The integrative capacity of a double term cannot be matched by most any other form of crystal except gridworks. This is because the double term allows interspacial linkage accuracy to the "nth" degree. It cannot be calculated with clarity on Earth. The quality of double terms to promote multiple and individual information receival is apparent in these forms. *(7th chakra and up)*

DIOPSIDE: Diopside is a calcium magnesium silicate. Polishing star diopsides bring some of the light out of the somewhat drab stone. For those needing a respite in a pelagic playground, diopside has water memories packed into it for you. It is excellent as a meditation tool and bathside companion, which will provide to the quiet time a sheltered cove of your choice. It is a conservative mineral, which helps people to sustain themselves through a tough work week or in traumatic situations. It is like an insurance policy against bad times. It can spread itself over a threatening situation and make things less overwhelming. It's money in the bank for a rainy day. Its character is very low-key, and it's just the mineral for hanging in the background until it's needed. This is an excellent mineral for maintaining digestive and nervous system health, and is recommended for wearing for colon cancer, ulcers, colitis, Crohn's disease, diverticulitis, arthritis, kidney stones, urethritis, excessive sweating, and improved hair and nail growth. It is good for the bones, teeth and heart, and for pregnant and lactating women. *(1st chakra)*

Chrome Diopside - Chrome diopside is diopside with chromium in it, and is colored a luscious green. Clear crystals make nice jewelry. Chrome diopside provides vibrations of deep relaxation. It harbors the weightless quality of floating in water, and sings a tangible note of serenity, stillness, and open-hearted trust. It is excellent energy for heart tissue and heart issues. *(4th chakra)*

DIOPTASE: As a copper silicate hydroxide, dioptase is perhaps the finest expression of copper silicate. The crystals never grow to enormous sizes; the size of the little finger nail being remarkable. It is called the "copper emerald." The fairies love dioptase because it embodies one of their favorite colors, and the crystalline structure is especially healing to their sorts of predicaments. People drawn to wear or use dioptase are fairy "dusters," meaning the fairy teachings have been a part of this person's past, and part of their present work is to provide a sense of the imaginative to others. There is much in this mineral regarding the need to be free and play so that one is not overcome by one's commitments to purpose and mission. It is a lighthearted healing mineral which promotes the expression of ecstasy on a level which brings satisfaction. Dioptase resonates to hummingbird, who is the willing transport bird of injured and wingless fairies.

Dioptase is easily polluted, and must be cleansed once a week to maintain its "amperage." Placing by water, following a moon exposure, is excellent. Never treat with salt water solution for cleansing. Dioptase is easily polluted, because it is so effective at removing negativity, stress, tension, toxins, and even shock (after natural or man-caused disasters) from living things, rather than environments. Dioptase can be combined with violets to produce excellent elixirs. Daffodil and dioptase elixirs can counteract depression and postpartum blues. Cranberry and dioptase elixirs create a joyous heart. White lily and dioptase is an excellent preparation for PMS sufferers. Dioptase in listed in the Spiritual Apothecary in the Auric Cleanser division. *(4th chakra)*

DIOPTASE and KINOITE: Kinoite is a hydrous calcium copper silicate, and this combination is usually from the Christmas Mine in Arizona where kinoite originates. This rare combination is like brushing your teeth. It cleanses and keeps tissues from "decay." It is an excellent combination for old people as it stimulates glands and hormones which tend to slow down in old age, and it brings back a strong recognition of divine play, which may have been lost during the adult years. It fosters a return to the sacred in an innocent context of trust. This combination heralds the presence of the fairy king and queen, because it brings in the divine spark of life which resonates to the fairy realm. See Dioptase and Kinoite as individual minerals. *(4th through 6th chakras)*

DOLOMITE: Dolomite is a calcium magnesium carbonate which is usually a white to creamy brown color, but the crystals can be pink and yellowish also. Dolomite has many uses industrially from building material and metallurgy to chemical applications. Metaphysically, it is considered the "herb" of rocks, because it is so useful as a tonic and aid to the body. Dolomite is a natural abrasive and toxic sponge in one. It stabilizes human and mineral energies, and it provides a buffer between disparate energies, allowing clear focus on each frequency. Strategically placed in a home with multiple minerals, dolomite creates balance in the atmosphere by removing stress frequencies. It can be thought of as a mineral of service and selflessness as it brings so many benefits wherever it is without drawing attention to itself. It is excellent for the nervous and endocrine systems. It helps bones, teeth, muscles and enzyme production. It helps digestion and lactation.

DRAVITE (Brown Tourmaline): See Tourmaline.

DUMORTIERITE (Blue Desert Lapis): Dumortierite is an aluminum borosilicate oxide. It is often a deep, dusty blue color. Sometimes found mixed with talc and muscovite, dumortierite is an excellent aid to past life recall and a releaser of old negativities. The talc with it is excellent for absorbing old charge out of the aura. The muscovite is a shamanic power mineral, but dumortierite qualifies for this classification on its own, because it contains boron. This combination enhances present life connection to times of knowledge and skill utilized in past life. One may experience a tubing or tunneling sensation when working with this mineral. This signals that dumortierite is helping to focus attention. It resonates strongly with heron energy, especially great blue heron. Conscious use of dumortierite signals a willingness to respond in awe to the beauty of the universe, a desire to praise, and humble acceptance of divine grace. Dumortierite is an Auric Opener in the Spiritual Apothecary. *(6th chakra)*

DUMORTIERITE with Quartz (Violite): This form of dumortierite is a reddish/brown, sometimes as crystals, on a white (quartz) matrix. It brings an awareness of ordination, receiving one's recognition as being qualified of the Source or the most divine. It is a service-oriented mineral resonating to the qualities of the number nine. It is an excellent stone for pastors, ministers, nuns, priets, monks, etc., or any individual serving humanity in the name of love, light, humility and charity. *(6th and 7th chakras)*

EDENITE IN QUARTZ (Madagascar): In this form, edenite is a green amphibole group mineral included in quartz. It is a hydrous sodium calcium magnesium aluminosilicate. Edenite is a powerful releaser of resentment and resistance. It increases flexible thinking, and promotes an environment of allowance and tolerance, especially after periods of grieving a loss. Its name encapsulates beautifully the innocent trust inherent in Eden symbology. It captures the inherent intelligence built into the universe and which sparkles out of all life. It redeems the role of serpent to bring wisdom, not just intelligence, to humans, because in ancient belief, serpent is a wise creature to be trusted and honored. The Hopi believe the best prayers are transmitted to Source in the presence of a snake over a three-day period. Edenite is deeply connected to Goddess energy represented by the serpent and its wise, healing powers. It is a powerful stone to utilize on an altar or in prayer. Like other amphibole group minerals, edenite resonates to frog energy. *(4th and 7th chakras)*

EILAT STONE: Eilat is a mixture of chrysocolla, azurite, malachite, cuprite, and turquoise found in the old King Solomon mines of Israel. The combined effect depends on the mix in the chunk being used. Since one of these copper-based turquoise minerals is usually present, eilat is an excellent harbinger of the coming of Spirit and Light. It enhances creativity, enlarges receptivity to prosperity, and facilitates openness to divine presence. *(1st, 4th to 6th chakras)*

EMERALD: Emerald is usually a dark green color in the beryl family. Beryls are beryllium aluminum silicates. Emerald is the physical plane embodiment of the Great Seal of the Serpent Mother. This green beryl can be likened to the kiss of death, the small death which brings detachment from the temptations, desires and weaknesses of physical life. Emerald is the great protectress of the Cave of Self where self-realization resides within. Emerald set in gold as a pendant or ring brings together the two great serpent energies in mineral and elemental forms. Emerald represents the Mother, the Earth, the great abiding receptive vessel while gold represents the masculine influx of solar energy. This very powerful combination may be difficult to wear continuously, and one's awareness of its ability to strip the self bare is essential. For removing baggage from one's existence in order to empty oneself for Source, there is no other more powerful pairing.

Emerald is a mineral which bypasses normal psychic function, bringing awareness directly to the heart where compassion rules. Since it brings so much to the individual, it enhances inner serenity and detachment from the karma of others. In one sense, many dysfunctional people can call this cold-blooded, a definite truth of the serpent. However, emerald stimulates courage to embody Light as a beacon rather than bleed from the heart endlessly for the trials and needs of others. Emerald empowers the knowledge of when to walk away and when to stay and teach others through example or resonance. Emerald improves all physical heart functions, as well as emotional functions of love and kindness to others. It calms the heart, and relieves guilt. Emerald activates the thymus gland to aid immune system development and function, and aids in the treatment of liver and lung diseases. It detoxifies the fourth chakra area of the body. Emerald provides qualities of endurance and protection to the entire trunk region of the body. Emerald is the state gem of North Carolina. Emerald is the May birthstone. Emerald is found in the Spiritual Apothecary with red beryl in the Etheric Sealer division. *(4th chakra)*

ENARGITE: Enargite is a copper arsenic sulfide. Enargite is a good friend to devas, sprites, fairies, animals and plants. It supplies a good energetic foundation for creating a fairy garden in the home. Dioptase can be cleansed by placing near enargite. Using these two minerals together often produces pleasing results — unexpected joyful happenings. *(1st chakra)*

ENSTATITE (Cat's Eye - Dark Green): Enstatite is an orthorhombic pyroxene, chemically called a magnesium silicate. It sometimes displays a star pattern when cut and polished into a cabochon. Metaphysically, it is called an Arcturean nekton. A nekton is usually associated with aquatic animals on earth who swim essentially free of wave and current action. The analogy being presented here is that enstatite is an excellent connector mineral from Earth to Arcturus, because it allows vibrational static to be neutralized so that communication is clear and vivid. This analogy holds true for making all manner of intuitive connections, not just to Arcturus. It aids maintaining a lubricated nervous system during psychic work. This mineral supports the teaching of Christ "to be in the world but not of the world." Estatite allows self-light to remain pure and true to its original conviction without being engulfed, dimmed or swallowed by other light or dark. It is an excellent companion stone to iolite. *(3rd and 4th chakras)*

ENTIGRITE: Entigrite is a third chakra equalizer, allowing the concepts of right use of will to be integrated gently into a busy, modern life. As we normally view our lives in Western culture, keeping up with modern life requires a high order of organization and task scheduling. However, entigrite deepens commitment to serving, nurturing and fulfilling self before attempting to address the myriad demands of our daily lives. This orientation ensures that we are at our optimal level to let our light shine and serve others. It is effective in small quantities for environmental impact, so meditation with small quantities often produces significant effects and results over a three-week period. Meditation cycles can be calibrated with the moon. *(3rd chakra)*

EOSPHORITE: Eosphorite is a hydrated manganese aluminum phosphate hydroxide. It forms in a series with childrenite. Often its short, brown, tabular crystals are associated with rose quartz crystals, from Brazil. This mineral is an excellent energetic aid to coagulation. It should not be used by those with high blood pressure or tendency to strokes or blood clots. It can be quite beneficial to hemophiliacs. *(1st chakra)*

EPIDOTE: Epidote is a calcium iron aluminum silicate hydroxide. It forms mainly in blocky, green, flat crystals or acicular fans. Epidote forms a series with clinozoisite and is classed with clinozoisite and zoisite in the epidote group. When a mirage surrounds the heart, it is as if the world to the individual is an indistinct, wavering concept which can be dreamed but never held in the palm of the hand. When man is cast into the sea of his personal doubt, he experiences the mirage, and he knows the irrational fear of reality displacement and conscious mental dissociation. This time in human history is incredibly trying, the survival of which is often accomplished not through wits and sensibility but through quiet, rest, sustenance, listening, prayer and collection of energies. Epidote is such a mineral, bringing to the sense of calamity a separation from reality, the cognition that rest is required. Epidote provides safety. Epidote heals and stimulates the linkage in the physical which allows information from other realms to be incorporated into conscious awareness. The individual is made more cognizant, has special or certain knowledge, often from firsthand sources (Webster's). This process is not easily elucidated. It is an enhancer of the personal process of power recognition and apprehension, in the sense of perceiving. *(2nd through 4th chakras)*

ERYTHRITE: Erythrite is a hydrated cobalt arsenate. The rich, velvety texture of erythrite hides its overall toxicity to living organisms, but interestingly, it provides counterintuitive support to overcome fertility issues.

Deep Red Crystalline - The "Queen's Mineral." Often called upon in ancient times to foster robust children, erythrite was well known to induce barren women to pregnancy or menstruation. It came to be called the Queen's Mineral because the need to bear an heir was often a huge responsibility which often interrupted menstruation, etc. A young queen was routinely given this mineral by her mother-in-law. The mineral was to be gazed upon into a quiet meditation or self-hypnosis. This mineral is also known for healing success in the male genitals as well. While the female suffered the blame for lack of children, the culprit was often the male. These highly-prized specimens come from Morocco. *(1st chakra)*

Pink (Hairlike) - Its bright crimson crystals appear in radial urchin-like form on matrix. Its bright patterns activate the lower chakras so blood flow improves for people with serious blood disorders like pernicious anemia. Holding between the thumb and forefinger of the left hand at the endjoint will infuse this energy into one's being, causing a healing reaction as the lower vital organs are stimulated. This form originates in Australia and Mexico. *(lst chakra)*

EUCLASE (Blue): Euclase is a beryllium aluminosilicate hydroxide most commonly found in Africa and Brazil. It is a highly prized gemstone. Metaphysically, it is the highest physical and auric protection mineral in the blue ray. It retards dysfunction. Over time, it supplies support to overcome dysfunctional impulses. It provides excellent, simple mineral energy for meditation, healing and psychic work. Like other beryllium minerals, euclase resonates to serpent energy, and offers auric protection resulting from accessing Goddess energies. Euclase is an Auric Sealer in the Spiritual Apothecary. *(6th chakra)*

EUDIALYTE: Eudialyte has a complex chemical formula which contains sodium, calcium, iron, manganese, zirconium and niobium. This is a rare mineral which exhibits a glassy, rich red color. It is found in a granite mix with calcite and augite in Canada. The passion, intensity and integrity of eudialyte is unequaled in the red, fuschia, magenta ray. It cleanses blood, and is a good companion stone to red clover and beets. It normalizes sex hormones, and supports strenghtening the immune system to counteract reproductive issues and sexually transmitted diseases. Eudialyte is a personal power stone, and deepens the ability to act appropriately regarding issues of power between people. Environmentally, eudialyte clears spaces of entities who have not passed on after death, and is a useful stone for counselors or practitioners working with terminal individuals. Eudialyte is the epitome of grace and joy in the sanctity of the heart. It is a mineral of circulation, therefore it aids all circulatory system parts and functions. It is a mineral of return and gratitude, bringing a distinct aura of tender appraisal to all situations. It allows one to see from the heart fully and without reservation or judgment. With arfvedsonite, as in specimens from Quebec, this mineral combination is a marvelous and profound teacher. *(1st through 4th chakras)*

EUDIALYTE with RIEBECKITE: This form of eudialyte is found in Alaska, and comes in small, red crystals spread through a matrix of white and black. Riebeckite is a hydrous sodium iron silicate. This mineral combination is a powerful integrator of the first, fourth and seventh chakras, and facilitates kundalini movement of energy through the spinal column. Its complexity addresses many energetic channels simultaneously allowing deep meditation with conscious stimulation. It is grounding, centering and uplifting. Environmentally, it is excellent for creative spaces where unconscious, conscious and physical aspects of being are stimulated together. It will enhance lucid dreaming. It is useful for people requiring grounding and centering as well as those lacking cardinality in their horoscope. Energetically, it provides a balanced masculine and feminine vibration. *(1st, 4th and 7th chakras)*

FELDSPAR: Feldspar is the group name for an important assemblage of rock forming aluminosilicates. Orthoclase and plagioclase minerals form the two major branches. The feldspar group includes albite, microcline and orthoclase. Jewelry-grade members of the feldspar/orthoclase group include moonstone, sunstone, spectralite, labradorite and amazonite. Feldspar is an excellent practical addition to any mineral apothecary. It is a team player and supporter. It can be likened to stage hands who change scenery, but do not perform on stage. Feldspar is excellent for preparing the stage for surgery, so to speak, and afterwards, comes in to clean up. If other cleansing devices are not available, run feldspar through layers of energy fields to ensure sanitary and purified conditions. Feldspar is an Auric and Etheric Cleanser in the Spiritual Apothecary. *(7th chakra)*

FETISHES: Fetishes are sacred animal stone carvings. The best known carving tradition in stone is done by the Zuni tribe of North America. Various carving materials provide their respective effects. In general, the impact of a carving is a sacred connection to the animal kingdom through spirit totem communion. Each carving tunes an individual to the powers of the animal and provides a specific and palpable connection for its user to the Earth community.

FLINT (Chert): Flint, a harder version of chert, is categorized as a chalcedony quartz, and forms as siliceous nodules in chalk or limestone sediments (Cliffs of Dover, England). This rock is comprised of zones of colors such as red, gray, blue, and black with white bands. As a member of the chalcedony vibration, flint is apotropaic, designed to avert evil. It helps to prevent nightmares. Flint aligns and strengthens meridians when mixed with oils and applied externally. It can be utilized as an overall tonic for health and strength. Occasionally, flint is found comprised of various minerals such as calcite, quartz, and hematite mixed with diatom and radiolite fossils. This produces varied effects, depending on the combination. Though used as a weapon, tool and fire starter for generations, the stone has great healing capacity, as it stimulates the regeneration of tissue throughout the organs. The predominant color of flint designates the greatest area of influence. Flint is the state gem of Ohio. *(Varies with color)*

FLUORICHTERITE: Fluorichterite is a sodium calcium magnesium fluorosilicate categorized in the amphibole group. It usually appears opaque in black to brownish, striated, tabular crystals, and is found usually in Ontario, Canada. It is a powerful healing tool with beneficial results in nearly all applications, something like the versatility of echinacea. The fluorine in its matrix repels bacteria, toxins, viruses, molds, fungi, parasites, and all manner of unwanted microbes. It can be an aid to fever in the body to kill off invading microbes. It is an excellent support mineral for allergenics and people with compromised immune systems. It helps in reaching conclusions in a difficult diagnosis, because it is especially attuned to energy pathways of the body. It is useful as an environmental stone in offices where work on the meridians of the body occurs, such as acupuncture. Like richterite, fluorichterite assists in recovery from loss caused by an enemy or negative presence in one's life. It is a mineral which promotes global awareness. Fluorichterite is listed in the Healers division of the Spiritual Apothecary. *(All chakras)*

FLUORITE: Fluorite is calcium fluoride. It is often found in pockets of shale in coal mines. Most commonly, fluorite is found in a range of colors from white or clear through purple, blue, green, yellow and golden. It can be flecked with pyrite, which is grounding and energizing. Chakra influence depends on the color(s) of the specimen. Fluorite brings higher states of purity into the physical plane. It balances mental vibrations, in particular, because it reduces negative energy in the atmosphere. Fluorite clears the mind of stress from fast-paced living. It is an excellent mineral for use in the work place, around computers, in meditation areas and healing rooms, because fluorite reduces static, friction and tension. Every room of the house, as well as public spaces, benefit from the presence of fluorite. It assists the physical body to actualize and open up to universal life force energy, and expands mental power. Cubic clusters symbolize advanced technological states in harmony with universal principles. The classic octahedron shape of fluorite's crystalline structure supports advancement of the individual in the modern world. In blue, purple, golden and white colors, fluorite affects the head and chakras above the head.

Fluorite requires no added measures for cleansing as it naturally repels negativity and toxins, acting as the "garlic" of minerals. Place specimens at a distance from amethyst as fluorite is so effective at clearing space, it reduces or tends to override the gentle, soothing vibrations of amethyst when they are close together. Fluorite strengthens the teeth and bones, and detoxifies the spleen, kidneys, appendix and gall bladder. Fluorite is the state mineral of Illinois. Because it does such a good job of clearing negativity and inducing safety, fluorite is placed in both the Etheric Cleanser and Opener categories of the Spiritual Apothecary. *(Whole body)*

Blue-Banded Opaque - Found in Peru and Mexico, this massive, harder-than-normal form of fluorite has lavender and blue bands, making it especially attractive and useful to individuals who work a great deal from the head, whether psychically, scientifically, or spiritually. It tends to provide a resolution energy which settles in after mind and brain field disturbances have been eased and released by its presence. It is applicable to migraine headaches. *(3rd, 5th, and 6th chakras)*

Rainbow (with Pyrite edges) - The "Philosopher's Stone." Rainbow fluorite with pyrite edges, found in China, facilitates grounding the physical body to the geomagnetic forces of Earth and elevating the upper chakras from heart to third eye. This allows clear receival and transmission of telepathic messages. The Philosopher's Stone was thought to be an imaginary stone, substance or chemical combination capable of transmuting baser metals into gold, and therefore sought by alchemists. See Webster's dictionary. However, the alchemical process activated by this special specimen is electrokinetic, meaning the shape shifting phenomenon practiced by certain adepts is aided and strengthened. This mineral combination is one of the finest on Earth. *(1st and 4th through 6th chakras)*

FORSTERITE: Forsterite is a magnesium iron silicate. It is an end member of the olivine group crystallizing in the orthorhombic pattern. It forms as tabular, striated, peridot-color crystals with a wedge-shaped termination. Forsterite cools the blood and body tissues when an individual is suffering in a hot climate. It is excellent for cooling down a tempermental, exciteable, angry and aggressive personality. It tends to stimulate dopamine production. Forsterite is centering. It brings attention to the heart chakra, and encourages cool decision making based on the wisdom of the heart rather than emotion of passion. *(4th chakra)*

FOSSILS: Though this is a wide subject to place in one category, it is worth mentioning. Fossil literally means "dug up," and is the word for mineralized remains of living organisms from past geological eras. Due to the strong life force that is preserved in fossils, holding fossils while meditating can be uplifting when this stored information is tapped. In addition to this, the material preserving and replacing the fossil has its own energetic impact. They also provide a natural link for those wishing to develop conscious bonding with archetypal aspects of plant, animals and humans.

For scientists, the greatest story ever told is preserved in the fossil record, and paleontologists seek to tell that story with as much drama, eloquence and willing suspension of disbelief as can be found in any movie theater or great religious book. Fossils help to identify the age of rocks widely separated from each other, because the succession of flora (plants) and fauna (animals) has been described and formalized. A great deal of information about the earth's crust has been obtained through the study of fossils or paleontology. Paleontology has three separate branches as may be predicted: paleobotany (plants), paleozoology (animals) and paleoanthropology (humans). Fossils justly can be called the record of living organism response to the life force or light energy. As described by Arthur M. Young, light is first cause. It is the organizing principle of universal life which cannot be separated into smaller pieces. He says there are seven stages to light, a prime concept of most spiritual traditions. Young says the cause of consciousness is light photons from which nuclear particles and atoms derive. Atoms crystallize into molecules (minerals), and light bounces off this firm foundation to return to itself first through plants, then animals, then humans. The total freedom of light descends into physical form and complete constraint to return again to enlightenment and total freedom through living organisms. Earth holds witness within herself of this progression from light to materiality to light through the fossil record. Fossils are the ancient library of how life responds to light.

Mineralization is the most common form of fossil preservation, and this involves mineral replacement of living parts or filling of spaces. Carbonizaton occurs in anaerobic or oxygen depleted environments where the organisms give up oxygen and nitrogen to bacteria leaving carbon-rich deposits of coal and oil. At 270 million years of age, a period was specifically named Carboniferous, in the Upper Paleozoic era, to describe the world's fossil energy resources. Fossil encrustation occurs when organisms are coated, leaving a negative impression, and distillation involves depletion of volatile elements, leaving a thin film of carbon showing the form of the organism. *(1st, grounding and bonding)*

Ammonite - The spiral form of ammonite, a cephalopod mollusk, was achieved by primitive forms of cephalopods by the Silurian period of the Lower Paleozoic era. It evolved into its own subclass called ammonoidea probably from the nautiloids. Its shell is called involute, because the last turn completely surrounds or enfolds all the other turns. Ammonites were highly successful during the height of their expansion. Ammonites probably traveled about in the same manner described for orthoceras. They completely disappeared at the end of the Cretaceous period, about 60 million years ago. In the end, they developed strange curling and twisting habits. Ammonites are an important guide fossil for dating paleontological zones of the Mesozoic era during their relatively short burst of fame. Ammonites are beautiful fossils, because they are decorated with ridges, turbercules and spines which form intricate spiral patterns. Ammonites also started out as straight-shelled animals. Ammonite fossils can be pyritized or covered with iridescent mother-of-pearl.

Brachiopod - Brachiopod is a small marine invertebrate which may be related to worms and bryozoa. They continue to live in marine environments as a community, but not a colony as with clams and muscles. Brachiopods shells are symmetrical, and shapes include: plates, ovals, hemispheres and globules. They have been found today at enormous depths (as much as 17,000 feet), and they are found to be successful in their ancient habitats as well. Most common today in the Mediterranean, Caribbean, south Pacific and Atlantic coast of Africa, brachiopod species number approximately 260 in about 70 genera. In the past, they reached 25,000 species, and made up a vast percentage of marine animals. They are superior guides to paleontological dating.

Coprolite - Coprolite is fossilized excrement often from amphibious dinosaurs, probably turtles. It is usually an iron oxide replacement of dung. See Coprolite.

Megalodon Tooth - Found in river sediments in South Carolina, fossil shark teeth are the remains of the ancient shark megalodon which could reach lengths of 80 feet. It is not uncommon for teeth to measure the size of a man's palm, but they can be quite small also. They are colored black to ocher or tan. Being very popular, and somewhat rare, fossil shark teeth can command high prices, if they are complete and undamaged.

Orthoceras - A primitive nautiloid, orthoceras (orthoceratides) is a 400-plus million-year-old Ordovician to Triassic period cephalopod mollusk originating in the Lower Paleozoic era, making it one of the oldest commonly found fossils. Most fossil shells are formed of aragonite which turns into calcite during fossilization of the hard body parts. Mollusks are highly useful to describe local geography at the time of their deposition, a study called paleographical reconstruction. Cephalopods exclusively are marine animals. Quite successful in their habitat, they are considered the most evolved mollusk. They are useful to age rocks, because they are widely distributed in sedimentary layers. Orthoceras is the most primitive form of nautiloid, because it is straight (orthoconic) or just slightly curved (cirtoconic). It could grow to six feet in length. It is the precursor to the ammonite which began to appear in the Silurian period and the present day Nautilus shell. Orthoceras which evolved into spiral shells could travel 1,600 vertical feet a day following plankton from its preferred habitat to the ocean's surface at night. They achieved this mobility through nitrogen-rich, gas-filled chambers which they could control in order to travel vertically.

Trilobite - Trilobite is a primitive marine arthropod that lived entirely within the Paleozoic era. The name derives from possessing three body parts of one long central section and two lateral parts. Trilobites often had many appendages, and they were capable of swimming and crawling both. They were well developed by the beginning of the Paleozoic era, and by the end of the Silurian period, they could roll themselves up like potato bugs. They disappeared by the end of the Permian period. Their flat nature seems to indicate that they lived on the sea floor like present-day king crab. There were species of trilobite which were adapted to water surface existence. Trilobite is another important guide fossil for stratigraphic analysis.

Walrus Tooth or Tusk (Fossilized) - Found in marine ledges in Alaska, fossilized walrus tooth or tusk is used for carving.

FRANKLINITE: Franklinite is a zinc manganese iron oxide. It is found with willemite in Franklin, New Jersey, and it is a classic fluorescent mineral. When crystallized, it has a spinel shape.

FUCHSITE (CHROMIUM MUSCOVITE): This silver and green mica family mineral resonates strongly to creative healing of feminine aspects such as earth, plants, the female id, and physical aspect. It is very powerful for herbal practitioners. It should be kept near the bed to influence the individual in subtle states of consciousness. *(4th chakra)*

GALENA: Galena is a lead sulfide, and serves as the source for native lead. It is silver colored. Galena's simple chemical formula and cubic structure provides balance and grounding to the etheric, emotional and physical bodies. It is an infection fighter as it contains sulfur, and strengthens metabolic processes which detoxify the body. It is excellent for grounding like other metals; however, galena is also adept at centering, reducing confusion, and bridging the higher self into the physical body. Galena is an excellent example of the tension between physical desires and spiritual aspirations, therefore it is useful for maintaining a steady momentum toward spiritual goals without losing contact with daily responsibilities and the joys of simple existence. It is an excellent mineral for dowsers. It is best used environmentally, especially around children, rather than worn (in a pouch) or handled, because of its lead content. Galena is the state mineral of Missouri and Wisconsin. *(2nd and 3rd chakras)*

GARNET: Garnets are a group of cubic silicates containing, in most common and accessible forms, one or several of either iron, calcium, aluminum, chromium or magnesium. The garnet crystal system is isometric, and garnets commonly form in dodecahedrons. They are usually found in mica schists. The durability and endurance factors of garnet make it a mineral which supports personal power and dignity. Especially in the red varieties, garnet enhances commitment and communion in love and unity. It promotes fealty and loyalty, and can be thought of as the classic stone of chivalry. It was thought to ward off physical harm, especially during the time of knights and holy wars. It is the best choice to represent the sign of Capricorn or the month of January, because it strengthens ease in responsibility and public affairs. It deepens familial bonds and encourages honoring elders and the aged. Deep colors of garnet cleanse the bladder, colon, kidneys, liver, intestines and stomach, and purify the blood while benefiting circulation. All garnets are Healers in the Spiritual Apothecary. Garnet is the state mineral of Connecticut and the state gem of New York. Star garnet is the state gem of Idaho. Garnet is the January birthstone.

Almandine (Red or Spessartine) - Almandine is iron aluminosilicate. In older times, it was called carbuncle. Red garnet stimulates blood flow and inspires love of life as well as survival instinct. Deep red almandine has the needed iron to cure anemia. Drawing energy from deep inside, creativity is inspired during meditation with this mineral. It enhances all areas of passion. *(1st chakra)*

Andradite - Andradite is calcium iron silicate, and is a garnet oriented to the region of the shoulders. It alleviates stress and tension in this area which in turn relieves pressure on the abdomen and colon, especially for women. This is because the color variation crosses the shoulders. Usually a muted golden brown, andradite provides support to people feeling overworked, exhausted and depleted. *(1st, 2nd 3rd and 5th chakras)*

Demantoid - Demantoid garnet is the household equivalent of aspirin. It has a sure-fire ability to address emotional distress and to provide pain relief. It became fashionable to wear demantoid garnet in past times, because so many people used it for pain relief for chronic ills which could not be cured. *(All chakras)*

Grossularite (Orange) - Stimulates the pituitary gland by aiding the production and secretion of growth hormone. Sit quietly with this mineral and observe a sensation of widening and relaxation of the skull and scalp between the ears and a stimulation of the spine and nerve endings into the feet. Imagine toxins flowing on the conduit into the Earth or back to the Light. Breathe deeply and release. *(5th chakra)*

Grossular (Massive Green/Pink) - Technically, this mineral is called hydrogrossular. It is from Transvaal, South Africa, and it is a calcium aluminosilicate. This form of garnet belongs to the family of stones which combine pink and green naturally. It enhances partnerships — whether personal, business or political. It facilitates listening and improved communications. For relationships which are long distance or experience frequent separations, it enhances a durable sense of togetherness and connection. *(4th chakra)*

Grossular (Massive Pink) - This stone is also the hydrogrossular form found in Transvaal, South Africa. It is a soft textured, massive garnet. The tenderness exhibited by this form of garnet is unique to its type, and facilitates acceptance, allowance, tolerance, and patience. It is useful to those with body image issues and eating disorders, such as bulimia, anorexia nervosa, or bingeing. It alleviates feelings of hysteria and calms the nervous system. It deepens a sense of safety in the jaws of the lion, trust that the universe will provide, and stopping eating at satisfaction, sufficiency and adequacy. Grossular garnet harmonizes so deeply with the child's need for comfort and nurture that it facilitates connections outside self which are healthy and wholesome. For those who have committed great wrongs against others, grossular garnet encourages change of heart, honesty and re-establishing new grounds for trustworthiness. This form of hydrogrossular garnet resonates to ant medicine. *(4th through 7th chakras)*

Grossular (Yellow/Green) - Has qualities of cleansing the digestive areas as the yellow/ green color works up from the typical first chakra influence. *(3rd chakra)*

Hessonite (Orange) - Similar effects as grossularite (orange). *(5th chakra)*

Melanite (Black, Andradite Group): This stone promotes persuasive abilities in people. It is a stone of positive speaking. It enables the wearer to project self-confidence, assurance and credibility. It is useful during job interviews or in careers where self-promotion is necessary. It combines well with hematite and biotite. *(1st chakra)*

Rhodalite (Pink/Purple) - This mineral is the stone of the seventh sister or seventh heaven. A woman born seventh in line is said to possess unequaled saint-like attributes of healing and salvation who will live a life of service to others which transforms the people she meets. Her presence has rippling effects out to the seventh generation and energy layer of life. This mineral aids in the attainment of detachment or bliss. It resonates strongly to the Himalayan region. *(Auric chakras)*

Uvarovite (Green) - Uvarovite is calcium chromium silicate, and forms in bright green drusy crystals which commonly grow quite small. Excellent specimens are found in Russia. A male-oriented version of dioptase can be found in uvarovite garnet. Again, it relates to the magical kingdoms, but rather than fairies, uvarovite garnet relates to gnomes and leprechauns, or the "little people" of a more masculine quality. In elixir form, it is excellent alone as an antidote to insect bites or as a natural laxative when combined with citrine and/or blue calcite. It produces a nice tincture when combined with mullein. Always put mineral in glass to separate it from tincture or elixir solution before activating. Uvarovite garnet is an excellent heart chakra healing mineral. It is especially sympathetic to gentle release of past heart trauma, followed immediately by infusions of green healing waves and heart energy acceptance. This is an excellent stone for healing practitioners. Since it is highly resistant to absorption of negativity, it can be used by or on a progression of individuals without constant cleansing. Once a month is plenty in sun and moon. It is a release enhancer. *(3rd and 4th chakras)*

GARNIERITE: Garnierite is a hydrous nickel silicate. It is a bright green, massive rock which is quite crumbly. Garnierite has similar properties metaphysically to smithsonite though they are not alike chemically. Garnierite also aligns the astral and emotional bodies, and like smithsonite is good for supporting the human electric and magnetic properties. Garnierite in particular strengthens balance and mental acuity, and can aid people who wake up slowly to be more present, when needed. Green. *(6th chakra)*

GASPEITE: Gaspeite is a nickel magnesium iron carbonate listed in the calcite group of minerals. It is a bright lime-green mineral from Australia. It facilitates transfer of chemical ions across cell membranes. It works well with herbs, tinctures, and acupressure/acupuncture. Gaspeite strengthens chlorophyll and photosynthesis, and aids utilization of chlorophyll in the body, especially for counteracting bad breath. It fights toxins, viruses, and bacteria. A stone of soldier-like preparedness, it is excellent for keeping the body strong during demanding and stressful circumstances. It is highly activating, so it helps people who have a hard time removing themselves from their couches. Gaspeite stimulates long-range planning, preparation for change and catastrophe, and moving out what is no longer necessary for survival or comfort. Gaspeite resonates to squirrel consciousness. *(3rd chakra)*

GEHLENITE: Gehlinite is a calcium aluminosilicate. Its color is brown. It acts as a hormone equalizer which balances mood swings and supports self-awareness to maintain presence and poise under stress. Gehlinite balances masculine and feminine energies. *(1st chakra)*

GEM SILICA: Gem silica is high quality chrysocolla. It expresses a clear and high blue color which impacts the fifth chakra, upgrading and enhancing communication abilities, while adding "compassion in action" to the wearer's skills. The yin energy of this stone makes it an emotional balancer and extends its influence to the heart chakra. This mineral truly impacts the soul, and is sought after for healing layout work. For those wishing to connect with the music of the spheres and the energy of St. Germaine, this mineral is a profound ally. Gem silica is rarely found alone, usually with better grades of chrysocolla. Gem silica is located in the Spiritual Apothecary in the Etheric Opener division. *(4th, 5th and 6th chakras)*

GIBBSITE: Gibbsite is an aluminum hydrate. It is occasionally a creamy brown accessory mineral to crocoite. It can look like clay. It's a very simple mineral with a simple job of enhancing structure and shielding. It provides a containment and protection awareness for the open receptivity of the feminine-based crocoite. On its own, gibbsite increases the sense of safety for those feeling vulnerable and exposed. *(1st chakra)*

GLASS (Fulgerite): In addition to the manmade variety, the cycles of nature make glass by fusing silicates and other minerals. It varies in color and shape since it is not crystalline. Besides the specific color healing which can be accomplished by back-lighting the mineral and taking 10- to 20-minute treatments (e.g., yellow glass cleanses the system of toxins), the mood of a meditation can also be enhanced, such as blue for serenity, violet for spirituality, etc. It is interesting to note the homogeneous and therefore universally androgynous nature of this substance. Some sources say glass can be of extraterrestrial origin, which would liken it to tektites. *(Varies with color)*

GLAUBERITE: Glauberite is a sodium calcium sulfate. It is often found as pseudomorphs after selenite. It is good for the garden as an antifungal agent. It is also good in combination with other garden purifiers, cleansers, and balancers. In conjunction with rose quartz, plants are able to metabolize available soil nutrients more efficiently, and produce larger, healthier blooms and fruits. Excellent to experiment with as it is very versatile and harmonious. *(7th chakra)*

GLENDONITE: Found on the Olenitsa River on the Kola Peninsula in Russia, glendonite is a calcite pseudomorph after ikaite, which is a calcium carbonate. It is a brick red ball of multiple dog's tooth crystals sticking out from a central growth point. It is noticeably heavy for a calcite mineral. It is interesting that this specimen emerges from a country which has struggled with the ideas of "to be rather than to seem" and "to be or not to be." Russians endure their long winters and the interminable expanse of Siberia with a kind of dogged determination the rest of the planet often fails to evaluate correctly. They call their land "Mother Russia," as it is a feminine place of stark reality, absorbed in long winters which shred the sensibilities of weaker souls. It's a place of night, cold and dark rather than dry, hot noontime. Russians shed their skins, emerging reformed yet remarkably recognizable, and they struggle with alcoholism and dark, moody creativity which threatens to eject them from stability, if not the planet at times. Glendonite is an apt metaphor for the Russian spirit and genius of reformation. *(1st chakra)*

GMELINITE: Gmelinite is a hydrated sodium calcium strontium potassium aluminosilicate. It is a creamy orange member of the zeolite group. It is a mineral tuned to the continuum of existence. Its power is to facilitate knowing the past and future as immediately as the present is felt and known. It is an excellent conduit for information flow, especially technical. It is receptive to alien transmissions as well as prophetic spiritual contacts. It regulates body fluid balance, and behaves like green leafy vegetables in the body. It provides support to exchange food and wastes in cells. It strengthens liver and muscle function. It balances over-acidity, helps the kidneys, and assists with weight control. *(Upper chakras)*

GOETHITE: Goethite is a hydrous iron oxide, which can be easily misidentified with limonite. It forms often in black, acicular crystals. It can be likened to blue/green tourmaline, because goethite resonates to the power and majesty of the ocean which offers humans an abiding sense of quietude and peace internally. The tidal rush of liquid earth rising from her crystalline core resonating with the fluid grace of water is given form and play in this mineral. For the true soul-yearning magic of this planet is contained in the mysterious truth which is water (blood), and to walk on water is to honor the life blood of this planet. For humans to experience clarity of purpose on this planet, they must be redeemed by water, for it is the element which calls them home, and which creates for humans the recognition of purity and the possibility of life. To be brought back to health without the use of and respect for water (blood) is not likely. Goethite is the water mineral. *(1st chakra)*

GOLD: Gold is an element which cannot be split into even smaller fundamental components. Gold is formed in hydrothermal deposits, and can be found in placers or alluvial deposits where host rock is weathering away, especially below waterfalls and rapids. Gold is called sectile, because it can be slivered without powdering by a knife, and it can be pounded to within one molecule of thinness. Gold is considered by many cultures to be the master stabilizer, harmonizer, balancer and grounder, and it was esteemed to prevent personal and spiritual corrosion. Gold is a symbol of wealth, prosperity, love, generosity, fertility and power. As an all-purpose healer, especially for the heart, gold is found in the Spiritual Apothecary in the Healer and Auric Sealer divisions. The Indians call gold "mineral light." Buddha is often depicted in gold to represent absolute perfection and enlightenment as well as spiritual freedom. Gold represents the sun, sacredness, and illumination in many cultures. In Western alchemy, living gold is the end product of working with the masculine and feminine properties of sulfur and mercury. In Eastern alchemy, gold is the cinnabar or essence of the heavens. Cinnabar is a mercury sulfide. Gold is associated with ouroboros, the Great World Serpent, who is the Lord of Gold, and it is associated with the Rainbow Serpent. Gold is said to harbor secret knowledge and the words of power. When these are ingested, the result is digested truth and light. Gold was thought to be the excrescence of gods, and dowsers say that the vibrations of gold and excrement are quite similar. Gold mining fever has provided blatant examples of human greed from the Cortez destruction of Mesoamerican culture to the rapid influx of European culture into the western United States. Minted gold was considered a spiritual abomination. Gold is the state mineral of Alaska and California. *(2nd through 4th chakras)*

GOLDSTONE: Also known as monkstone, goldstone was invented by Italian monks and kept a well-guarded secret for several centuries. Goldstone is unusual, because the copper tetrahedrons should melt in its host glass, as the melting point of glass is higher than copper. Instead, the tetrahedrons maintain their integrity and produce a sparkly character very pleasing to visual contact. Since goldstone is copper silicate, with coloring agent added to the glass for effect, it resonates with dioptase, and is a massive alternative to dioptase for regular wear and constant contact. Goldstone is made into carvings, beads, donuts, and cabochons making it highly accessible for perpetual impact.

Gold - Goldstone, like diamond, is a good deflector of unwanted energies, and is highly regarded in the spirit realm as a protection mineral. It is a nervous system stimulant. It enhances masculine traits and the transmission of healing energies from the hands. Therefore, it is applicable to long-distance healing, Reiki, chi gung, cranial-sacral therapy, massage, chiropractic, animal training, gardening, cooking, and many other uses. *(2nd and all chakras)*

Blue/Purple - Purple goldstone is an excellent decorative mineral for protecting psychic and empathic sensitivities. It can be used most effectively in a public setting with excellent results, although it is not as powerful as snowflake obsidian for truly difficult circumstances. Purple goldstone resonates to female energies, and is effective in transmission of healing energies from the heart, throat and third eye. It is also applicable to long-distance healing, as well as hypnotherapy, yoga, dance, contacting the deceased, public speaking, singing, chanting and toning. *(6th and all chakras)*

GOSHENITE (White Beryl): Goshenite is the clear or white variety of beryl, which is beryllium aluminum silicate. Goshenite is a fabulous mineral with classic vibrations. It aids the receptivity of sublime vision. It fits into every crowd and brings to every event or moment a sense of dignity, power and grace. Goshenite captures the imagination of humans in utilizing the broad spectrum of mineral energies, and it aids all humans and minerals in reaching their vibratory excellence. Goshenite pierces the veil of maya or illusion which separates the ego and analytical mind from true recognition of reality. Goshenite stimulates deep meditation, hypnosis and dreams. It can be thought of as a dream weaver. Goshenite prompts the body to embrace etheric energies, as it resonates to both air and fire elements. Goshenite embodies the flying form of dragon as well as dragonfly. In the Spiritual Apothecary, goshenite is an Etheric Opener, and makes way for healing energies to be received, incorporated and released as completed visions of power. *(7th chakra)*

GRANITE: Granite is a medium to fine grain igneous rock which is the most common rock worldwide on the surface of continents. It is an igneous rock which often cools very slowly. The alkaline feldspar variety of granite must contain microcline or orthoclase and quartz. Accessory minerals within it are often plagioclase, biotite, pyrite, zircon, tourmaline, and magnetite. A secondary variety of this type will not contain plagioclase. To replace that mineral, amphiboles and sodic pyroxenes are present with accessories of fluorite and rare earths and radioactive minerals. The potassic feldspar variety of granite must contain quartz, potassic feldspar (orthoclase, microcline), plagioclase and biotite mica. Accessories to this type are magnetite, apatite, pyrite, zircon and tourmaline. Occasionally, muscovite mica, garnet, hornblende and pyroxene will be present. Granite's matrix contains the same properties of clear quartz in a more dispersed pattern with the grounding effects of biotite mica and pyrite, the first chakra affects of garnet, and specific chakra affects of the color of tourmaline which is present. It has been used historically on a massive scale as a building stone. It is a good foundation stone in building structures to balance energies of a place. Granite is an unusually low-key, yet efficient stone. When combined with feldspar, it bridges the first, fourth and seventh chakras, supplying an even more effective foundational energy. As with other prevalently seen stones in our everyday world, we must remember not to take this rock for "granite." Granite is the state stone of New Hampshire, blue granite is the state stone of South Carolina, and red granite is the state stone of Wisconsin. *(1st and 7th chakras, or 1st, 4th and 7th chakras)*

GRAPHITE: Graphite is one of two forms of crystalline carbon. Diamond is the other form. Graphite is black, soft and flaky. Masses of graphite are up to five percent silica. This form of carbon has many practical uses, but is not typically thought of as a healing mineral in the spirit realm. It is best used as a mineral's mineral, meaning that it serves the needs of other minerals. Place it in proximity to the place where healing minerals are stored between uses, and allow it to cleanse them.

GROUTITE: Groutite is a hydrous manganese oxide. This rare black mineral is a collector's prize. Groutite enables release from insomnia. It should not be used by people who sleep well and deeply, as it will produce grogginess and sluggishness. It will aid in deepening hypnosis as well as trance states, if desired. *(1st chakra)*

GYPSUM: Gypsum is calcium sulfate with water molecules which grows in a clear, tabular form most frequently in twins shaped like spearheads. Selenite is a form of gypsum which can be found in characteristic fishtail or swallowtail shapes. Satin spar is fibrous assemblages of elongated satin-textured crystals. Alabaster is an occasionally banded form of gypsum which appears waxy and granular. Gypsum is highly soluble in water, especially warm and hot water, so it should never be soaked or left in the rain for cleansing. Gypsum responds well to sound cleansing with bells, gongs, chanting and singing. Gypsum is the basis of sheetrock or wall board in construction, and it is mined for plaster of Paris, retarder in Portland cement, fertilizer, and as a flux for pottery.

Desert Rose - Desert roses grow in rosette shapes, often incorporating sand into their structure, creating a reddish-brown color.

Golden - The lack of density in the weight of gypsum makes it conducive to space travel and work. This means that for those attracted to concepts of extraterrestrials, spaceships, and the like, this mineral provides primarily two aspects: (1) the wearer sends out a frequency of alert yet friendly interest to the cosmos. It acts something like a radio transmitter. For those with malicious intent, it zeroes the frequency to ground and the individual becomes a non-conductor; and (2) for those of psychic expression, it provides a stronger awareness of link-up, something like adding a stronger beam to the frequency so interaction between earth and space is enhanced. Holding in the left hand increases receival, and holding in the right hand increases transmission. Holding in both hands allows the psychic to optimize both. It can be worn on the body, and it can be worn over the third eye to free the hands for psychic work. *(7th chakra)*

Selenite (White) - Selenite is not recognized as a formal mineral name. It is a form of gypsum which is generally clear or white and tabular with distinct cleavage layers. See listing under Selenite.

HALITE: Halite is sodium chloride or salt which crystallizes in the cubic system in a "hopper growth." Salt precipitates readily in land-locked bodies of water with high salinity and briny conditions. It often associates with anhydrite and gypsum in Permian and Triassic rocks around the world. In the Salina beds of the Silurian Age and overlaying Mississippian Age, extensive salt deposits occur in the northeastern United States. It was so vital to health in ancient times that it was more valuable than precious metals and gemstones. In alchemy, salt represents the physical and astral bodies. Alchemists believed that salt fixed the volatile spirit, and was the necessary static element between mercury and sulfur. Salt is applicable in many cultures to physical, moral and spiritual transformation. Salt was considered apotropaic by the Romans, and was placed on the lips of infants for eight days after birth to dispel negativity and evil. Greeks, Hebrews and Muslims viewed the sharing of salt as a sign and seal of friendship, and salt was an important part of sharing bread. The Hebrews also revered salt as a symbol of spiritual discernment. The Japanese spread salt daily on thresholds and throughout the house after the visit of an unwelcome or detested person. Salt is an important aspect of everyday life for the Japanese and is used in Shinto for cleansing.

Supersaturated solutions of salt can be left in rooms overnight in glass jars and then disposed of only by toilet to cleanse negativity, disease and evil from rooms. Salt has many important associations to physical health. Overabundant use of table salt results in overweight conditions, edema, arthritis, and conjestive heart failure, so obtaining sodium and chlorine from vegetable sources provides the body the correct balance for health. Food sources of chlorine and sodium combine to cleanse the body and blood, remove waste matter and reduce fat. As an energetic support in the environment or worn in a protective cloth, halite helps to maintain bacteria-free and sterile conditions. It helps to reduce insect infestations. It is a good mineral specimen to keep in the kitchen, preferably under glass. Combined with calcite or a mineral with calcite in its chemical structure, halite reduces swelling and edema, helps heart function, and asssists with eating awareness.

Halite is recommended for healing work, especially as a permanent fixture in the room or under the table for similar reasons as using fluorite. It is best used on the body for severe infections, cancer, HIV/AIDS, herpes, shingles, entities, paranoia, and dissociation. Otherwise, having it near the body is sufficient. Halite requires no cleansing, because it resists negativity, toxins, and all living organisms which are harmful to humans. All forms of halite are found in the Etheric Cleanser and Healer categories of the Spiritual Apothecary. *(Whole body)*

Blue (Purple mix) - Fibrous seams of this color occur in France. Blue halite, especially with purple in it, is an excellent soother for people disturbed by their psychic abilities or mental problems such as obsession/compulsion. It is generally good for drawing negative energies from schizophrenics as well. It is excellent for headaches caused by organic toxins, parasites and living viruses and bacteria. It is excellent to use in any counseling environment. *(6th chakra)*

Pink - The pink variety is often found crystallized in dry salt lakes of California. Recognizing the importance of salt to life on earth, in ancient times, pink halite was called, "the Great Sister," and was used by sisterhoods ceremonially as well as medicinally. It is listed in the Spiritual Apothecary under Etheric Cleansers, and was usually the first mineral given a young woman for that category when she began her instruction as a healer. As a medicine, it was used energetically. For severe illnesses, patients would lick the mineral or be given small amounts dissolved in tea. As an environmental piece, pink halite cleanses the room and immediately begins the process of etheric detoxification as people enter the room. *(Etheric chakras)*

Rock Salt - A time-worn substance in the sense that it is of the Earth and is a good toner of Earth beings. But the substance is also a good emissary of Earth consciousness to those who are new to Earth or are returning to Earth after a long sojourn elsewhere. The substance is like the burnt embers of a fire picked up by the wind, then deposited in a trickling stream. It draws Earth elements together. It is union. It is speechless, yet its intent is understood. So no language is required, no teaching must be given. It will calm agitation, heart palpitations, anxiety. It provides a link to Earth vibrations which may be missing for some. It is not a grounder like obsidian. Instead, it is more diffuse, bridging across elements rather than drawing down to one in particular. In clear or pink normally, but interaction with heat or other minerals colors it black, brown, yellow, blue, or purple. *(2nd and 3rd chakras)*

Yellow/Orange - This cubic or stalagtitic variety is found in New South Wales, Australia.

HANKSITE: Hanksite is a chlorinated potassium sodium sulfur carbonate. It forms in cloudy, opaque, yellow crystals which exhibit a "Herkimer diamond" shape, and often is included with earthy tones. It looks like a lemon drop. The greatest impact of hanksite is to bridge between the second and third chakras. These little gems help open up the "energy conduit" between these lower centers for healing and personal empowerment. It also stimulates high spiritual mentality with its subtle golden hue. The double-terminated aspect, which is a trademark of hanksite, naturally helps the energy flow between affected areas. Chemically, hanksite provides the perfect blend to support moving heavy foods such as meats and grains through the digestive system. It counteracts all disorders associated with regulating body fluids, such as high blood pressure, overweight, fluid retention, etc. It restores liver and kidney function and alkaline conditions in the body, therefore it helps counteract arthritis, rheumatism, and gout. This mineral is a major infection fighter from skin to muscles to bones. This mineral is the perfect example of eating green, leafy vegetables, from dandelions to kale to mustard greens. It can be called the comfrey of minerals as it provides excellent poultice support. *(2nd, 3rd, and 9th chakras)*

HEDENBERGITE: Hedenbergite is a calcium iron silicate and is a member of the pyroxene group, which makes it a relative of jadeite. It is a pressure release stone for the shoulder area, where the weight of burdens may be carried, because it draws focus deep into the first chakra. Energy reserves can be accessed and the blood stream is aerated, increasing stamina and mental efficiency. It is a good mineral for body work applications all along the spine and pelvis where it provides connection between the nervous system, blood vessels, organs and vertebrae. Stress often robs the spine and pelvis of calcium, so healing applications of hedenbergite help to counteract this depletion. It aids osteoporosis and anemia, and brings color to the face and skin. It normalizes acid-alkaline balance and supports the nervous system. Hedenbergite is recommended for releasing and working through grief complicated by guilt and feelings of responsibility. It can aid in recognizing one's true scope of involvement with loss or death. *(1st through 5th chakras)*

HELIODOR: Heliodor is a yellow to light green beryl, which is beryllium aluminum silicate. It crystallizes in pencil shapes with etched surfaces. It is primarily found in Brazil and Afghanistan. Heliodor is a high choice to use in healing applications to prepare the physical and subtle bodies for energy infusions, because it cleanses the intake mechanism of the third chakra, where humans take in their inspiration for life. It possesses resonance to both the water and earth elements. Heliodor brings in the energy of serpent sunning itself on a rock. It is a life giving embodiment of nature at its most sacred. In its terrestrial aspect, heliodor blends well with schorl to ground auric and etheric energies and release toxins while healing occurs. As a jewelry stone, heliodor continually removes unwanted energies. With longterm wear or exposures, it increases expressing oneself as honest, blunt and direct. Heliodor resonates to dolphin energy, and in its resonance to water, it brings in the magical and cleansing qualities of water. Heliodor is listed in the Spiritual Apothecary under Auric Cleansers.

Gold/Yellow - Gold heliodor aids in restoring good relations in business associations which have gone awry. It also reinvigorates personal intentions to achieve goals which seem unattainable. Gold heliodor is an excellent stone to wear for people who are highly involved and committed to tense and tight schedules. Heliodor is the stone of the Delphic oracle, and is to be viewed as bridging the soul with the intellect and the psychic center of the third chakra. It stimulates ecstatic as well as lucid prophecy. As a temple stone, it helped to induce temporary states of trance possession which allowed the oracular voice to come through. It is the harbinger of Manitu, the Great Spirit of the Algonquin tribe. As such, it prepares the physical body for transmission of food from spirit, the energy of kundalini. Heliodor embodies goddess energy of Aphrodite, a water goddess. Like other beryls, heliodor blends and works well with other minerals. It stimulates chakras above the crown as well as the third chakra. In the Spiritual Apothecary, heliodor is found in the Auric Cleansers category where it cleanses the subtle and physical bodies prior to being opened for healing energies. *(3rd and upper chakras)*

Green - Heliodor is an excellent mineral for supporting intention to forgive or make amends with others. The green vibration especially relates to matters of trust in love or to close love relationships. Green heliodor also strengthens feelings of self-esteem. It helps to overcome shame and guilt. It unleashes joy. Green heliodor resonates to the playful aspect of dolphin, and increases the sense of belonging and community. *(3rd and 4th chakras)*

HEMATITE: Hematite is iron oxide. It crystallizes in blades, and can grow massively or in botryoidal veins. Hematite looks like a silvery, black mirror when polished, cut or faceted. Hematite is grounding and stabilizing. It resonates to bone marrow and red blood cells, and is applicable to health issues stemming from bone marrow. It aids blood aeration, anemia, memory and mental fatigue. This is the best mineral to support the respiratory system, as it aids oxygen movement in the body while providing a stable foundation for the emotions, which often trigger respiratory distress. The silvery, black surface is much like an early camera. It has a photographic memory, because its crystalline structure provides enormous templates for records to be made. When placed in a situation, hematite will record what occurs around it, and the scenes can be viewed as if reeling through a movie. Once all the many interstices are full, hematite must be cleansed or erased before it can record anything new. Sun and moon for a week is suggested, plus mental applications of deprogramming the crystal to accept new information. Hematite can be utilized as a witness stone in healings which release all forms of trauma whenever experienced, such as past life, birth, childhood or crisis situations. The hematite can absorb much negativity and release it later. When polished, hematite deflects negativity. Hematite is the state mineral of Alabama. *(All chakras, especially the 1st)*

Rainbow - Rainbow or specular hematite is an excellent chakra system balancer, toner and stabilizer. It provides biofeedback regarding centering and integration of the subtle body systems with the physical body system. Excellent mineral for subtle body system practitioners and all physical/subtle body energy movement practices such as tai chi, gin shin do, chi gung, reiki, etc. It provides grounding as well as opening to electromagnetic impulses. *(All chakras)*

HEMIMORPHITE (Calamine): Hemimorphite is a hydrous zinc silicate which can bear a beautiful sky blue tone. Its name implies the meaning of being half-formed. It induces thought/nervous system integration. When a person is in a coma, it will aid the reintegration of brain circuitry and chemical functioning. The neurons and electromagnetic charge of the brain are exceptionally susceptible to the healing quality of this mineral as its near-form state is what is needed to overcome some brain dysfunctions caused by accidental blows, etc. It is not so good in addressing hemorrhage, for example. It is conducive to beginning the proper integration of the mind in dyslexia. This stone, combined with amethyst, smoky quartz and selenite, will reduce hypertension and hyperactivity caused by chemical imbalance and fluctuations. It is an extremely useful stone for children until the age of 12 when the delta waves begin functioning in the brain. Hemimorphite relieves mental stress and anxiety for children in difficult situations. In light blue botryoidal to white tabular crystals, it works on the head region. *(6th to 7th chakras)*

HERDERITE: Herderite is a calcium beryllium phosphate which usually forms in beige, "sheavy" masses. It tends toward a white shade with occasionally green undertones. It is very useful to humans as a mineral which induces full body awareness. "I just knew it my bones" is a succinct description of its applicability to humans. It connects humans to the serpent energy of earth. Herderite is unequaled in the process of earthprinting. Earthprinting is the art of determining the best locales for individuals to settle. Herderite allows the evaluation of physical needs such as climate and food types to be melded into the evaluation of cultural, social and spiritual needs. We may not have been born or live in the optimum place to achieve a centered, peaceful state of being. Herderite is also sensitive to earth energy lines, and is useful to geomancers, geopathologists, dowsers and feng shui practitioners. It is an excellent choice to combine with iolite for many physical issues. See Iolite. *(1st chakra)*

HERKIMER "DIAMOND" QUARTZ: This particular variety of high quality, clear quartz, usually double terminated, is found in Herkimer County, New York. This unusually pristine species of quartz formed in pockets of dolomite-rich rock which repelled the quartz, allowing the nodules to grow suspended in solution rather than on a host rock, which is common for other forms of quartz. The diamond-like appearance of its natural faces looks like facets of a cut stone and is actually the result of a high percentage of unimpeded growth. Herkimers open and enlarge the aura and higher energy bodies, plus the physical body, allowing high frequencies to enter. When used for clear energy transmission, they enhance communication with higher frequencies. Herkimer diamonds are recommended for use over the third eye at night for dream work or during meditation to enhance higher self, spirit guide, angelic and archangelic connection. They can also be placed under the pillow. When placed at the throat during healing sessions, Herkimer diamond smooths and regulates the flow of white light energy into the lower body. It can also be held between the fingertips while meditating. Herkimers are excellent for wearing after intense spirit bonding through dream work or meditation, as they will enhance constant connection to the Light within. Herkimer is located in the Spiritual Apothecary in the Auric Opener division. (6th and 7th chakras)

HEUBNERITE: Heubnerite (or hubnerite) is a manganese tungstate. With iron, it is called ferberite. They form a solid solution series with each other. Heubnerite forms in deep red, striated and heavy crystal blades. It is often found associated with clear quartz from Brazil. Heubnerite provides deep connection to the classical period of history, but on a global scale so one is not confined to Greece and Rome. Its heavy quality supports deep meditation. It is important not to use this mineral in water as the relaxation may be too complete and there could be risk of drowning. This mineral can help those who have trouble being hypnotized to be relaxed enough to achieve release. *(1st chakra)*

HEULANDITE: Heulandite bears four possible expressions, depending on the concentration of calcium, potassium, sodium or strontium present. It is a hydrated aluminum silicate which contains calcium, potassium and sodium, and in two types it contains strontium and/or magnesium. It is classified in the zeolite group, and usually exhibits silver/gray to white fan-like blades. It can have red, orange, brown or green inclusions depending on the host rock and associated minerals. Heulandite is a mineral relating to beings originating from Sirius. As taught there, it aids in healing a dysfunction of speech called dysphasia. It was often so effective that people healed rapidly and became well known speakers and those who recite sacred texts. Women were especially prone to the speech healing qualities of heulandite. Heulandite is an excellent meditation mineral. *(5th and 6th chakras)*

HIDDENITE: See Spodumene. Hiddenite is green spodumene.

HOPEITE: Hopeite is a hydrated zinc phosphate. The rusty red crystal clusters show a triangular hollow nature. For people with foundation issues, the inability to feel safe and settled, hopeite is a wonderful companion and friend. It vibrates to the number four, bringing security and second awareness to our surroundings. It is an excellent mineral for aviaries, zoos, and places of human confinement.

HORNBLENDE: Hornblende is a complex silicate classified in the amphibole group. It forms in a complex solid solution series with several other amphibole minerals. It contains calcium, sodium, magensium, iron and aluminum. It is an important member of metamorphic rocks, but it can be found in igneous rocks. Its black crystals can be stubby and double terminated. Hornblende is the ambassador or mediator of crystals, promoting the manifestation of brotherhood through allowance and acceptance of differences rather than through enforcement of pseudo like-mindedness. It tends to promote the expression of personal qualities, thereby showing individual uniqueness. It is a mineral capable of reducing the emotional static of tension and conflict between parties attempting to reach agreement on an issue. It helps individuals to reach clarity on what they perceive as their vested interests, and to accept compromise so that both parties can achieve resolution. Hornblende contains aspects of grounding, and is a highly stable interactor with other minerals and humans. Hornblende helps to highlight what was special about loss to provide it recognition, and to then motivate one in concrete ways to begin moving beyond grief and sorrow. *(1st through 4th chakras)*

HOWLITE: Howlite is a hydrous calcium beryllium silicate. It is usually a white stone, unless dyed, with gray lines running through it. It is found in cauliflower-like nodules. It is commonly used in jewelry. Howlite is a very important mineral metaphysically, but is usually not recognized for its higher attributes. It is a member of the serpent group of minerals. It cleanses the aura, aids in purifying blood, and activates Higher Self and guide contact. It protects the chakras, particularly upper ones and also chakras of the astral body. It brings energy (higher levels of light) to the wearer. Water energized by this mineral is very beneficial to the human system. It aids in removing phlegm when worn near the throat. Companion plant is sage. Howlite is located in the Spiritual Apothecary in the Etheric Sealer and Auric Cleanser divisions. *(7th chakra and upper)*

IDOCRASE: See Vesuvianite. Idocrase is the old timer's name for this hydrated silicate of lime. It usually forms in blocky patches of green and black in a massive stone. Idocrase (Vesuvianite) is listed in the Healers category of the Spiritual Apothecary.

Massive - Bears the properties of the colors of vesuvianite listed under vesuvianite. Vesuvianite is usually crystallized.

ILVAITE: Ilvaite is a calcium double-iron silicate hydroxide which forms in short, black, square, and shiny crystals. Ilvaite is readily useful to all blood and cellular processes. It counteracts anemia, arthritis and gout. It stimulates the first and second chakras, creative processes, and activities requiring manual dexterity and focus. It helps to build stamina and endurance. It tends to burn energy, so the individual needs to keep track of blood sugar levels while utilizing ilvaite. Cleanse the body with water or smudge prior to use. *(7th chakra)*

INESITE: Inesite is a hydrated calcium manganese silicate hydroxide. It forms as a pink coating in pockets, and is most commonly found in Trinity County, California. Curiosity and exploratory tendencies are emphasized by inesite. It is an excellent mineral to use with students who have begun to lose the fervor to learn. It enables listening as a form of absorbing information, but also it encourages direct experience through hands-on participation or exposure. It is a masculinized version of manganocalcite, because inesite has silica in it. *(5th chakra)*

IOLITE: Iolite is a more common name for cordierite, which is a magnesium aluminum silicate found in India. It is also called water sapphire. Iolite is a color change mineral shifting from translucent blue or violet to brown depending on the angle of light refraction. This stone is an opener of the head region from the throat to the two chakras above the crown. It enhances auric penetration for healing work by drawing the lavender ray deep into the root chakra. Because of this ability, it is unequaled as a discernment stone which draws "other" senses into consciousness. Utilizing its energy regularly deepens the ability to feel true/false readings at the heart level. It is an excellent stone to support energy practitioners, such as reiki, chi gung, and gin shin do. Iolite is an excellent resource stone for magnesium to support bone health, but more importantly, magnesium regulates the endocrine system and the nerve fibers of the central nervous system. To gain maximum physical benefit from iolite, it must be accompanied by a phosphate mineral such as amblygonite, brazilianite, herderite, lazulite, purpurite, turquoise, variscite or vivianite. A combination of the two minerals will address mental issues from retardation to mental illness as well as muscle cramping, so it is excellent for athletes to combine these minerals as well as intuitives, empaths, psychics and anyone needing mental health support. This combination tends to support alkaline conditions in the body as well as elimination of waste products. This combination helps to replace important elements in the body after alcohol consumption. It is valuable for electrical issues, sterility, tension, hypothyroidism, inflammation due to acidity, and enzyme function. For plant essence practitioners, this combination behaves much like valerian when calcite is added to the mix. This combination of three mineral types is effective for chronic fatigue. Whenever the above symptoms are severe, add calcite to the combination for perfect support. Iolite is listed in the Auric Opener division of the Spiritual Apothecary. *(5th to 9th chakras)*

IRON: Iron is a metallic element, and in conjunction with cobalt, nickel, vanadium and manganese, iron forms steel which is the backbone of the Industrial Revolution and the base of modern civilization. It is a hard worker which combines well with titanium and molybdenum for high technology uses which allow humans to excavate stone on a scale never before achieved and to fly. Its early use in the Industrial Revolution for trains and cars has been superseded by aluminum, however iron remains a vital constituent of life. It is important to note that the crystalline core of Earth is iron. Iron is essential for hemoglobin and must be catalyzed by a trace of copper and chlorophyll. It transports oxygen to cells and helps to fight off bacteria and viruses. Iron addresses all bleeding and bruising issues of soft tissue. Iron is necessary for respiratory enzymes and to counteract anemia.

Good sources of mineral specimen iron are hematite, pyrite, marcasite, arsenopyrite, limonite, goethite, siderite, ilmenite, iron meteorites and sphalerite. Human artifacts of iron such as railroad spikes can be strategically placed around the house for energetic support in feng shui and grounding energies. Preparing food in iron cookware, especially tomato sauce dishes, provides dietary iron, because the iron reacts to the acid in tomatoes. Iron is often characterized negatively, such as the Hindu belief that the kali yuga, the age of iron and darkness (the age we live in now), brings in the fourth and final era in the cycles of manifestation. It is the metal associated with the planet Mars, death, destruction and war. In some cultures, women wore iron necklaces, probably because iron is especially useful to gestating women to ensure the fetus is receiving adequate iron for its circulation and blood needs. *(1st chakra)*

JADE: Jade is a common name for several minerals including jadeite, nephrite and tremolite. Jade is one of the toughest rocks known to man, because the structure is "knitted" together so well, making it difficult to cut and polish. It brings emotional relief at any stage of grief, loss, mourning, emotional crisis and pain. Choosing any jade to calm the heart chakra and bring emotional relief at any stage of grief and trauma is tantamount to triage and emergency intervention. Despite the opinions of ancient lore, all jades emit a feminine vibration of safety and healing energy. Jades are the epitome of "She Who Watches." The quality of detachment provided by jade goes a long way to explain why eastern culture has developed the forms of spirituality it has.

Jade is fairly accessible in Alaska, Canada and the Pacific Northwest of the United States, an area Native Americans acknowledge as "the place where healers come to be healed." The Pacific Northwest also is the region regarded as the pinnacle of development of Native American justice on the North American continent. Jade is appropriately revered as a stone of wisdom, longevity and luck. In China, jade literally represents the yang principle, because of its tough and indestructible quality. It is the Chinese symbol of the five transcendent virtues of benignity, lucidity, resonance, immutability and purity. Jade represented the ruler or king, and sending it to another was an indication of the Mandate of Heaven. Jade was thought to be created by lightning. The Mayan culture also highly regarded jade. It is thought to enhance immune system function as it promotes coherent body functions and communication between body/mind. Jade is the state gem of Alaska. Jade is an Auric Cleanser and Healer in the Spiritual Apothecary. See Richterite.

Black Jade - Black jade can be found in the Yellowstone region of Wyoming. Black jade is a master grounder, especially for people with a lot of air and fire tendencies or signs in their horoscope. Black jade is the sorcerer's stone, for either males or females. It is particularly useful for supporting light workers who are harassed by dark entities. It supplies power and confidence to stand firm in the face of darkness. It represents energies of the Mother Protector and Bear Mother. *(1st chakra)*

Blue Nephrite- This variety is found in California. Blue jade influences the brain, especially to counteract amnesia and senility. It is a good companion mineral for overcoming addictions, especially those stemming from a psychological origin, though it does provide mental relief for physical addictions. Blue jade used in conjunction with meditative practice supports chronic pain relief. It also serves to alleviate anxiety and panic attacks when worn continuously. *(5th and 6th chakras)*

Bowenite - Bowenite, or new jade, is a recent find in Korea, and it is light lime green and quite translucent compared to other forms of jade. Bowenite assists with deprogramming mental confusion or rigidity caused by religious or political persuasion and mind control. It aids releasing fear of loss of control and enormous waves of repressed emotional power. Bowenite stimulates healing grief over loss of identity and being subsumed into annihilation of the spirit. In turn, bowenite restores a sense of safety in self, safety in the company of others, and safety in letting go of dogmatic attempts to instill certainty of belief and predictability in behavior. *(4th chakra)*

Green Jadeite - Helps connect for dream work, causing spiritual learning with physical plane impact. Can assist drawing on past life wisdom especially related to the Oriental traditions. It has a therapeutic affect on many glands, organs and the circulatory system. In its various colors, it stimulates the first through sixth chakras. Jade projects universal love. Good to wear or carry. Jadeite is especially helpful in supporting physical expression of grief through crying, wailing, moaning and keening. *(All, 4th through 6th chakras primarily)*

Jadeite - Jadeite is a sodium aluminum iron silicate classified in the pyroxene group with spodumene, augite, diopside and enstatite. Jadeite is the highest gem grade of precious jade. It acts as an emotional release tool during dreams. It is of value when used in the sleep area or when placed on the throat. *(5th chakra)*

Lavender Jadeite - In the higher latitudes, sunset can often be suffused by layers of purple, lavender, indigo and blue light. Like this time of transition filled with high vibration, lavender jade is a stone which propels the individual into states and energies of magical intensity. For best results, it should not be worn continuously or every day. It must be properly honored through ritual respect. If so, it provides excellent support for times of committed spiritual interaction. *(6th and 7th chakras)*

Lilac Jadeite - Lilac jadeite, from Turkey, is an immune system booster and is useful in counteracting AIDS and HIV, reducing inflammation, and providing an overall aura of tranquillity and acceptance. Lilac jadeite used in conjunction with green and black jade is more effective, as this combination supports first and fourth chakra integration with the seventh chakra being stimulated by the lilac tone. The jade family of mineralization, used as a unit, tends to be conducive to female or feminine issues impacting the nervous and immune systems, while the tourmaline family of mineralization, used as a unit, more efficiently addresses male or masculine aspects of AIDS/HIV and immune and nervous system dysfunction. *(6th chakra)*

Nephrite - Nephrite is actinolite, which is a calcium magnesium iron silicate classified in the amphibole group. It forms a series with tremolite and ferro-actinolite. A deep, rich green, nephrite is unsurpassed as a gentle soother of emotional pain, especially fear, sorrow and grief. It also calms anger disturbances, and instills an ability to listen and hear both one's own internal voice and the voices of spirit guides, allies, and the higher self. Nephrite combined with pearl provides a cleansing quality which encourages letting go of resentment, bitterness, judgment, and patterns of thinking which encourage repetitious experiences of past trauma or abuse. Nephrite jades are excellent for use in meditation or sleep to support detachment, letting go, and raising the awareness of the desire to move on to new phases of life once grief and mourning have occurred. Nephrite jade resonates with frog energy. Nephrite jade is the state gem of Wyoming. *(4th chakra)*

Pink - "Starpath jade" (pink) is a stone which promotes remembrance of one's heritage and birth in the stars. It is excellent for starseed souls being born today. It is a nice way for star children to recognize each other. *(7th chakra and above)*

White (Pure Tremolite) - Tremolite is a hydrous calcium magnesium silicate classified in the amphibole group. White jade is often suffused with other tones which apply to the chakra affected by those tones.

For instance, white jade from Russia is suffused with a golden brown, which brings light to the second and third chakras. Other jades can be a green/white suffusion, emphasizing lightening effects on the heart chakra. In Eastern tradition, white jade is synonymous with immortality and the Philosopher's Stone of Eastern alchemy. It is a symbol of grace, charity and faith. See Tremolite.

JAMESONITE: Jamesonite contains sulfur, which is not bound in a sulfate or sulfide ion, as well as lead, iron and antimony. It is found in black masses of needles in Zacatecas, Mexico. The word diakinesis describes a stage in cell division which jamesonite facilitates. It also supports deep healing work which impacts DNA, allowing individuals to correct genetic abnormalities, especially related to blood, causing illness. It is a blood cleanser. Jamesonite should be held by the practitioner, rather than placed on the body, so healing energies are transmitted specifically. *(All chakras)*

JAROSITE: Jarosite is a hydrous potassium iron sulfate found in the alunite group. Jarosite can be of interest to geneticists as it facilitates cross-fertilization in plants and flowers. Gardeners can utilize jarosite to promote successful fruitation. It aids the regulation of fluid balance, and is an excellent option to increase fertility in males as this combination of elements increases semen, which carries sperm, as well as protein production for higher sperm counts. It enhances prostate health and cleansing of reproductive organs in both sexes. It is applicable to fluid retention issues such as overweight, edema and high blood pressure.

JASPER: Jasper is an opaque, massive, fine-grained, microcrystalline variety of chalcedony quartz with significant quantities of admixed material, frequently iron-oxides. Red is the most notable color in jaspers, but browns, blacks, greens and blues can be found, and color mixtures often create scenes. Generally, jaspers promote healing, tissue regeneration, and stimulation of the immune system. The placement and effect varies by color. Yellow jasper stimulates the solar plexus area. Red is used for bronchial problems and the lower chakras. Brown jasper grounds, and green jasper has overall healing use when worn, held in the hands, and placed environmentally. Jasper is thought to be particularly applicable to women, fertility and childbirth. *(Varies with color)*

Biggs - Biggs jasper is an excellent aid to hypnosis, as it relaxes the conscious mind while elevating dominance of the subconscious. It is a good tool for relaxation, behavior modification, stress relief, anesthesia and sedation. Its effects should be immediately apparent. If not, the individual is not susceptible to this mineral vibration and many find psilomelane a good alternative. *(1st and 7th chakras)*

Brecciated (Red) - Brecciated means the material was broken geologically and then healed. In this case, the healing was done by white quartz which formed in veins. This stone stimulates water absorption in cells, and restores normal fluid tension in cells after dehydration or trauma. It should not be used or worn by hemophiliacs or pregnant women. However, this mineral is beneficial to the hemophiliac after excessive bleeding has been stopped, and to the post partum woman to help restore hormone balance. *(1st and 2nd chakras)*

Bruneau - Bruneau jasper is a two-tone brown nodule with concentric circles which look like clouds. Bruneau jasper embodies the inhale of life force and the exhale of creative manifestation. A wonderful jasper for individuals out of balance in the second and third chakras. Enhances communion with personal totem animals. This is an excellent jasper variety for the gall bladder, spleen and appendix. *(2nd and 3rd chakra)*

Butterfly - Butterfly jasper is colored a creamy beige with yellow spots and squiggles of quartz. Butterfly jasper soothes the solar plexus chakra and strengthens the immune system by facilitating moderating mechanisms in behavior. If an individual is highly reactive to life stresses, especially minor ones, butterfly jasper helps to temper thought patterns which constantly stimulate the immune system into "fight or flight" responses. Being constantly aggravated by waiting in lines or at lights or being cut off in traffic is an indication that the body will eventually revolt and cause a tremendous need for rest and relaxation. Butterfly jasper, when worked with purposefully, will stimulate changes in behavior. *(3rd chakra)*

Hickoryite - Hickoryite is a reddish and creamy yellow two-toned jasper. It is a motivator. It gets the holder moving. It is a plan implementation stone. It builds up energy. It combats fatigue. Its primary function is at the physical level. It brings good luck. *(1st through 3rd chakras)*

Imperial, Green - Of the green minerals, imperial jasper relates to the head region, in particular the third eye, but also promotes integration of the chakras from solar plexus to crown. It is a good stone for hair barrettes, combs, tiaras and brooches, or bolos worn at the neck. It is best worn so that it can be seen by others, as it promotes integration with others. *(3rd through 7th chakras)*

Imperial, Red - Red imperial jasper relates primarily to the heart chakra, but also enhances integration of the chakras from throat to base. This is an excellent stone for rings, bracelets or belts because the hand moves frequently between these chakras and wearing at the waist strengthens the energy flow from throat through base. Again, is best worn visibly like green imperial jasper. *(5th chakra down)*

Juniper Blue - Juniper blue jasper has several applications to the ear, nose and throat area including easing symptoms of tinnitus and tonsillitis. It reduces teeth grinding, chronic hiccups, and nervous salivating or swallowing. Wearing juniper blue jasper at the throat area is not necessary; however, placing a piece in a rolled up scarf and worn on the neck will increase effectiveness. It enhances the sense of smell, and helps to relieve addictions. *(5th chakra)*

Leopardskin - Leopardskin jasper is colored by red, yellow and brown spots. Centuries ago, before any luminescence traveled the face of Earth, there were dark beings who lived all alone. It was not their habit to dwell in dark places, so they sent a messenger to the great Leopard Numi. Numi was black with flashing eyes, the only light on this dark orb at night. When the messenger reached Numi, he asked the Holy One to intercede with Great Spirit that there may be light to balance the lives of the dark beings. Dark beings have great weight and importance in the universe. Without their influence, there would be no galaxies or solar systems. Numi wanted to know if Earth's dark beings were ready to be pushed aside by the flashy, showy luminescent beings. Numi asked if the dark beings fully understood what it means to be important, critically important, yet disparaged, ignored or even vilified by light beings. "You have not seen how dark beings are treated in other areas of the universe. Are you ready for what will happen?" The messenger did not feel qualified to make such an everlasting decision yet he understood his mission and answered yes. The war between light and dark began then because light is blinded by itself and dark is blind without light. Numi asked Earth to form herself some leopardskin jasper as an antidote to luminescence fever. And what this means is that the universe has yet to reveal her deepest essence and mysteries. Judge not and you shall not be judged. What you do not know will be revealed. *(1st, 4th and 7th chakras)*

Mt. Tahoma Picture - Mt. Tahoma picture jasper is found in seams east of Mt. Rainier. The gold-yellow aspect is a visual aid, good for eye afflictions. The gray-black aspect helps tone the gums. Stimulates hearing or auditory abilities and sensitivities. All these aspects in combination help purify the intake mechanisms of the body thereby enhancing the quality of internal functioning and attunement. Much affinity to the awareness of Atlantean past lives. *(3rd through 6th chakras)*

Orbicular - "Messenger or Herald stone." Orbicular jasper shows a characteristic brick red texture with black spots. A classic location to find orbicular jasper is at Rialto Beach on the north side of the outflow of the Hoh River near La Push, Washington. The sending of orbicular jasper between heads of tribes or any individuals was an indication of the need to meet in person or call together the people. Used by a shaman in sacred ceremony, orbicular jasper represented the history of the ages as recorded in universal records. Orbicular jasper thins blood and enlarges capillaries allowing increased blood flow in the brain and tissues of the nervous system. An excellent stone for supporting the health of the placenta during pregnancy. It is second only to bloodstone in blood cleansing abilities, and strengthens hemoglobin. Orbicular jasper is a good companion stone when using red clover as a blood cleanser, and may be preferred by healers over bloodstone in some instances where bloodstone may not integrate as well as orbicular jasper with other remedies being used simultaneously. *(1st chakra)*

Pigeon Blood - This distinctive translucent jasper with red spots and plume areas is found in the petrified dinosaur fields of Utah. Pigeon blood jasper strengthens the will to face mother/mothering issues which can be expressed as miscarriages, menstrual cramps, irregular menstruation, endometriosis, violence toward women, rape and female exploitation. With conscious work around fears of mother, mothering, relationship to mother and the mother principle, pigeon blood jasper can provide a soothing support. It is also an earth (mother) chosen blending of the creative principle colors of red and white. It activates the meridians of the left thumb (white) and index (red) fingers. *(1st and 7th chakras)*

Rainforest (Green Rhyolite, Australia) - This jasper contains dots of white crystallization with agate areas which are a clear golden brown. The green jasper matrix also contains golden brown ribbons. Its expression is complex and changes rapidly in a small space. It combines energies in a swirling jumble which is quite pleasing. This form of jasper is excellent for liver cleansing and regeneration. It enlarges internal screens or palettes of awareness, allowing those who are private and reclusive to widen their perceptive field and cope with the presence of others either in their private or work space. *(2nd, 3rd, 4th and 7th chakras)*

Snakeskin - A red, white, black pattern stone found in Australia. This mineral combination is an energy mover, good to use during massage or energy work. It opens up the surface layers of skin and allows release of unwanted energies. It often will relax the brain allowing the sympathetic nervous system to also relax which enhances secretion of toxins out of the muscles. *(1st to 7th chakras)*

Spider Web - Even by its name, spider web jasper speaks to the interconnectedness of universal energies and processes. It supports attunement to awareness of how these energies and processes work, change and integrate to create the web of life. It connects two feminine chakras (the second and fourth) for compassionate creativity. *(2nd and 4th chakras)*

Sweet Home - This brecciated red jasper is from Oregon. Sweet Home jasper stimulates ovulation and fertility, and is an excellent support stone for the healing of reproductive organ diseases, also in men. *(1st chakra)*

Vistatite - Vistatite is a rare two-tone jasper of green and blue/gray colors. It is a very nice agate for connecting with water spirits. It is an excellent choice for aquaria and ponds.

Willow Creek - Willow Creek jasper provides scenic qualities in pastel yellow tones. It is good in homes or work places for people who need Earth connections and grounding. The process of connection is very cellular, having qualities akin to herbal extracts in the body, so it is mild if necessary, yet strong and invasive if necessary. *(1st chakra)*

JET (Lignite): This black mineral is a type of coal formed from carbonized wood. It has been revered as a protector against illness and personal attack by ancient civilizations, especially those of the European continent. People were even buried with it to protect them in the afterlife. It was used specifically for toothache and protection from the evil eye by the Greeks. It is a favorite carving material for the Native American Zuni carvers, because it is soft and easily worked. It is softer than petrified wood, so it is absorptive of negative entities troubling an individual, and can be cleansed with dry salt, soil burial or sun exposure. It was often a first choice for medicine pouches due its lightness and effectiveness as a protective stone. *(1st chakra)*

KAMMERERITE: Kammererite forms in deep magenta, fuzzy coatings. It is a female power mineral of exceptional quality. It aids in the recall of female past lives. It is a "shelter" stone, early recognized for its harmonic to the void, womb, cave. As woman is the conduit of whole life, she alone has been granted the sublime awareness of healing through the ability to harbor life. *(1st and 6th chakras)*

KERNITE: Kernite is a sodium borate. It looks somewhat like fibrous selenite. Kernite is the principal ore of boron found in Kern County, California. Greater than 95 percent of the world supply of boron is located there. Kernite enhances clairaudience and clairsentience. *(7th chakra)*

KIDWELLITE: Kidwellite is a sodium iron phosphate. It forms as a lime green coating with black aspects. Kidwellite is an excellent mineral to consider for Cushing's disease, also called Cushing's syndrome, which is a pituitary disorder often resulting in obesity and muscular weakness. *(5th and 6th chakras)*

KIMBERLITE: Kimberlite is the parent rock of diamonds, and is found in South Africa, Russia, Australia, New Zealand and Austria. It is an ultramafic rock that must contain clinopyroxene and pyrope (garnet). Accessories of diamond, corundum and kyanite are found. While kimberlite is an indication of diamond presence, it is an exceptional occurrence. Kimberlite is brecciated and gray from South Africa. As the host to so much beauty and greatness, kimberlite tends to pale in comparison and significance, however without the lowly manger, there can be no exalted birth. Kimberlite makes no claims. It just is. Simplicity has its rewards, because whenever kimberlite is noticed, the potential for stupendous discovery is greatly increased. Being content with one's lot in life has its rewards.

KINOITE: Kinoite is a hydrous calcium copper silicate. It forms as a light blue drusy. Kinoite is an excellent mineral choice to enhance spiritual discipline for the adept as well as the novice. It is a soothing, supportive mineral which allows those pastoring to others to maintain a sense of strong alliance with Source. It provides a direct conduit of divine awareness to the physical level, hence the statement, "I felt it in my bones," is an accurate description of how kinoite supports intuitive analysis. It is good to verbalize with this mineral as kinoite is a fine listener. Therefore, it is applicable to oratory and speaking the prophetic word. It is a powerful aid to skeletal, bone and teeth issues. *(5th chakra and upper)*

KOTTIGITE: Kottigite is a hydrated zinc arsenate, and a rare member of the vivianite group. Kottigite lessens sound irritations caused by tinnitus or sinusitis. Kottigite is an excellent mineral for throat ailments as well as earaches. It is an excellent stone for clairaudients. As with all members of the vivianite group, kottigite enhances a form of communication. In this case, the form of communication is receptivity to sound, listening and paying attention. *(6th chakra)*

KUNZITE: See Spodumene. Kunzite is the common name for pink spodumene. All colors of kunzite (spodumene) are located in the Healers category of the Spiritual Apothecary.

KYANITE: Kyanite is an aluminum silicate. Kyanite is trimorphic with andalusite and sillimanite, and all three minerals belong to a metaphysical category which is best described as, "I'll believe it when I see it." It is listed in the Spiritual Apothecary in the Auric Cleanser category. See Sillimanite and Andalusite.

Black - This form of kyanite is not grounding, because kyanite doesn't possess the mineral structure to truly ground. However, the black expression of kyanite empowers manifestation of vision and clairvoyance in real time. It can be a powerful tool of immediacy and responsiveness to thoughtform, helping to bridge highest thoughts to manifested reality. It pushes visionary ability deep within the self to be harnessed by physical abilities and discipline. Therefore, it is useful to any person who places images on paper or in computer. It is excellent to use during meditation and hypnosis or while listening to subliminal tapes. *(1st and 2nd chakras)*

Blue - Blue kyanite is an elevator, raising vibrations to an openness not accustomed by the user. It does not relate directly to alleviating depression and depression state symptoms, but it provides a higher level of awareness where the personality may look out and see the "light of day." This ability to rise above one's self-imposed vision of the world provides for the user a road map, an idea of what is available should the customary mindset be put aside for a moment. For those already attuned to an optimistic frame of mind and positive visualization, blue kyanite provides a lateral or horizontal widening out which gives breadth to vision, allowing the user to incorporate new areas of thought into the mental framework, thereby enlarging the scope of the individual's positive influence. In the aged, it provides a remembrance of youthful zeal and innocent idealism, thereby refurbishing familiar patterns of positive intent, and reducing the pull toward de-energized thinking which allows the onset of the death spiral. It is best kept wrapped lightly in cheesecloth or cotton rather than fragmenting the delicate structure through handling. *(6th chakra)*

Green - "Heart of the Lionness." This aluminum silicate in the green ray allows deep meditation to draw on the loving and healing qualities of the fourth chakra. This ray alleviates depression through seeing the light at the end of the tunnel, and stimulates the incorporation of dream work and visions into daily life. Its energy is feminine, and may be more attractive to those wishing to support receptive qualities in themselves as well as female visioning power. Green kyanite complements andalusite's ability to bring in global vision and local manifestation. *(4th chakra)*

LABRADORITE: Labradorite is a dark iridescent feldspar much like spectralite in appearance, but sunstone is a closer relative. Original specimens were from Labrador, but today, most labradorite originates from Madagascar. Night vision and flight close to the ground in dreams are enhanced by labradorite. It sensitizes and enhances heat sensing and electrosensitive abilities of humans and other animals. This supports detection of how close objects are at night, seeing auras, and imminent onset of seizures and earthquakes. Labradorite encourages development to the next higher octave, which is the ability to accurately gauge long distance healing frequencies being sent and received. Labradoite is an excellent stone for blind people. Labradorite sparks shift in consciousness through being able to see what is there, but could not be seen, because it was not understood in the context of one's current reality. It is said when the Spanish conquistadores arrived in the New World, at first, their ships were indistinguishable to the indigenous people who could only see the boats being rowed to shore from the ships anchored farther out. Labradorite encourages making transition from the ground to the air, from unaware of energy to aware of energy movement, from internal blindness to internal sight. It is an excellent companion stone to aurichalcite. Labradorite is listed in the Auric Conditioner category of the Spiritual Apothecary. *(6th chakra)*

LANARITE: Lanarite is the old-timer's name for this rock from Texas. It is brown with blue opal flecks. It is called "Mother's Helper" metaphysically. Soothes and comforts infants and children. Excellent for when children are ill or fretting. Provides grounding and security. *(1st chakra)*

LAPIS LACE: Lapis lace is a blue and tan rock from Turkey. This lazulite/calcite mix transduces sound vibrations to visual images. It is a profound mineral for psychic work involving clairaudience and psychometry, as the fifth chakra controls both as they connect to the brain. It is a powerful mineral for all sound healing practices as well as singers, toners, chanters and storytellers. It aids the narrow fifth chakra in passing energies through it whether in grounding of psychic flow or release of toxins and excess healing energies. Lapis lace is an excellent choice for cosmotron spheres. *(5th chakra)*

LAPIS LAZULI: Lapis means mixture, and lapis lazuli is a mixture of lazurite, calcite and pyrite. The chemical structure is fairly complex containing sodium, calcium, aluminum, silicon, oxygen, sulfur, chlorine and hydrogen. It is commonly included with white calcite or golden pyrite; both help integrate indigo energy of lazurite into the body. It fluoresces green under ultraviolet light. The prime locations for lapis mining are Afghanistan and Chile. Mining conditions in Afghanistan are as primitive today as they were 2,000 years ago, where men dig high on a mountain side and carry out the material in backpacks. It is more than a week's walk past the closest town to the mining site. Lapis is a popular stone for jewelry; however, it is excellent for environmental and feng shui applications. It projects a masculine vibration which is strong and protective. For people who require or who desire a more feminine vibration, please refer to sodalite, a lazurite relative. Lapis encourages mental calming while increasing psychic ability, and enhances spiritual study and growth. Its deep indigo tone relaxes the third eye for meditation and intuition support, and relaxes the eyes to relieve headaches and eye problems. It calms the brain to enhance restful sleep, and encourages deep, even and regular breathing to overcome sleep apnea and morning asthma congestion. It assists understanding interpersonal relationships by bringing thoughts to focus on self and the behaviors of self, and encourages one to see the self clearly. It has beneficial impacts on depression, anxiety, panic, compulsive disorders, catatonia, coma and memory loss. Lapis was used extensively in ancient Egypt. It was used as a symbol of the night sky, bridging the cosmos and earth in cultures from the Middle East to South America. Lapis lazuli is categorized in the Etheric Sealers division of the Spiritual Apothecary, because it is so powerfully protective and purifying. *(6th chakra)*

LAPIS NEVADA: "The Atlas Stone." Lapis Nevada is a thulite, diopside skarn. It lightens the burdens an individual may carry. Aids breathing. Opens up the back area very nicely prior to healing work or massage therapy. Excellent for back injuries. *(4th chakra)*

LARIMAR: Larimar is an ocean-blue form of pectolite, which is normally white. Pectolite is a sodium calcium silicate. Larimar can be striated with red dendrites, and the color tone can vary from green to light blue with white pectolite mixing all through. It was found in the Dominican Republic, which remains to date the only mining location. Pectolite is a hydrous sodium calcium silicate. In the spirit realm, larimar is called incandescent, because its energy is so brilliant and clear on the other side. It literally acts as a beacon which attracts good intentions, prayers, blessings and healing energy to the wearer. It is good for curing the common cold or coryza, respiratory flu and sinus infection. Of the blue ray, larimar captures the essence of blue flame, the point where flame burns hottest. It helps to burn out germs, viruses, bacteria, fungi and molds. Larimar provides internal fire the proper stage for expression, allowing all residuals of toxins to be emitted safely and healthfully. It is helpful for lung cancer, asthma, emphysema, and thyroid and lymph cancer. Larimar works well in conjunction with herbs, especially fireweed and coltsfoot. Larimar promotes the externalization of internal beauty and intensity. Larimar is found in Etheric Cleansers category of the Spiritual Apothecary. *(5th chakra)*

LAUREL LEAF FOSSIL: "Transcription Stone." There is a deep link to eternal principles in this fossil and it fosters the realization of the power of the Word as it was inscribed upon quartz tablets for Moses to reveal to the Chosen. The golden quality of this fossil gives a high level of protection to the stone and it is, therefore, excellent for amulets and other protection uses. See Quartz (Tabular). *(1st and 6th chakras)*

LAZULITE: Lazulite is a hydrous magnesium aluminum phosphate. Lazulite is a powerful mineral for the environment and can be used for healing applications on the bare skin for 10 to 20 minutes to reduce boils, sunstroke and inflamed insect bites. It can be placed near burn areas, but not on them, as direct application "freezes" the injured area too quickly and leads to moisture loss, inhibiting natural healing. Environmental benefits include promoting a calm atmosphere among usually tense individuals. It is especially conducive to promoting agreement during mediation of disputes. This is an excellent mineral for addressing all nerve and nervous system disorders and issues. It tends to work more beneficially in conjunction with calcite and a balanced nutrition, especially paying attention to vitamins. Enzyme production is enhanced as well as digestion and absorption of nutrients as well as converting food into substances the body can utilize. It can help with skin and fingernail health. It is important for the immune system and fighting viruses and bacteria. The qualities are especially adaptable to male energy, whether in the male or female body. Its deep color promotes tranquillity. Note that lazulite is complementary to iolite. Lazulite is found in the Etheric Cleansers division of the Spiritual Apothecary. See Richterite. *(All, especially 6th chakra)*

LAZURITE: Lazurite has a fairly complex chemical formula, which is best described as a hydrous silicate of sodium and aluminum with some sulfur and chlorine. It is a dark blue crystalline mineral sometimes called ultramarine, and is the basic ingredient in massive lapis lazuli. Lazurite is capable of smoothing out complexity into a simple and unified whole usually in a way which is pleasing to all involved. For instance, imagine placing lazurite or a mineral rich in lazurite in the middle of a round table where combatants have gathered to negotiate a peaceful resolution to their conflict. Everyone is asked to sit quietly, to gaze upon the crystal, and to naturally allow the eyes to close for a while, and breath. Can you imagine? Combatants closing their eyes in front of each other? Breathing? Love thine enemy. Love thy neighbor. So simply said, so difficult to do. As easy as shutting one's eyes and feeling the connection around, across, and through the table. Then upon opening the eyes to re-engage with the room, each other and the table, the combatants are asked to take equal leadership, equal ownership of the whole, the space that is defined by their presence. And they are asked to return, and return, equal to the vision and the whole which they uphold between them. The beauty of lazurite is that it provides impeccable clarity and purity in communication, leaving one with no doubt why it has been so highly regarded historically by royalty and other political leaders. It is a must gemstone for diplomats, ambassadors, consuls and officials of state. *(6th chakra)*

LEAD: Lead is a humble and useful metallic element which is poisonous to humans, especially children, in excessive amounts. The most notable lead-bearing mineral is galena. Fine fluorite specimens are often extracted from lead mines. Rome utilized lead for water pipes and placed it in wine, which led to disastrous consequences. Some buildings with lead roofs have lasted over 1,400 years without repair. Lead became popular for leaded windows, because it is quite malleable and easy to mold. See Galena. *(1st chakra)*

LEGRANDITE: Legrandite is a zinc arsenate. It is found rarely in crystallized, striated pencils in Mexico. It is golden yellow. It is a good aura cleanser, and is useful for counteracting anemia, and for rebuilding red blood cell levels. Legrandite resonates deeply with the solar plexus chakra. Legrandite is found in the Spiritual Apothecary in the Auric Cleanser category. *(3rd chakra)*

LEPIDOCROCITE: Lepidocrocite is an iron oxide, red needle inclusion found in some Brazilian and Madagascar quartz. It is like rutilated strawberry quartz, but has less dense needles. Lepidocrocite is a mineral of resolve, definitive recognition, and lasting choice. It empowers humans to stay the course and fulfill commitments. Because it is almost always found included in quartz, it brings in the ability to cognize with the ability to do and make real. It strengthens body awareness. *(1st chakra)*

LEPIDOLITE: Lepidolite is a mica containing substantial amounts of lithium. It is listed in the Healers division of the Spiritual Apothecary.

Lavender - A soothing and gentle relaxer for the environment. An ideal sleep mate for the bedside table or headboard area. This stone is very subtle, harmonizing well with energies and vibrations of other minerals, and provides them a home and is not changed by the associations; but those coming in contact who are susceptible to the carried mineral can absorb it without negative reactions. Its violet hue softly works on the third chakra to remove negative forms, and in turn balances emotional energies. Often a deep pink rubellite (pink tourmaline) is found embedded in the matrix. Lepidolite in located in the Spiritual Apothecary in the Healers division. See Lithium. *(3rd to 6th chakras)*

Yellow/Lemon - Yellow lepidolite is a rare variety found in the sacred Black Hills of South Dakota in conjunction with rose quartz and black tourmaline. This mineral triplet worn together brings grounding with insight into and acceptance of one's calling in life. It resonates with the third chakra and strengthens the bond with the higher self. It illuminates life purpose, and brings an enlarged scope of possibility to personal karma or mission. Yellow lepidolite brings awareness of past life history and the chain of purpose which links all lives into a cohesive total self. *(3rd and 7th chakras)*

LIBETHENITE: Libethenite is a copper phosphate which forms in drusy coatings of green to blue. The light seafoam-green aspect of this mineral resonates to old oceanic energies, and is good as a meditation focus to promote ocean healing and health. The darker blue aspect of libethenite is a more global representative of Earth energies which speaks to our ancestors from the stars. It calls up Star Woman, the bringer of life and genetic formation, and asks for her blessings to bring healing and health to Earth. *(4th through 6th chakras)*

LIBYAN BLUE AGATE: See Chalcedony.

LIMESTONE: Limestone can form chemically (called carbonatic) or from the accumulation of calcareous skeletons of marine life (called organogenetic). There are two types of organogenetic limestones. Fossiliferous limestone preserves specimens of organisms, and compact limestone is a fine grained compaction of calcite with occasional large specimens preserved. This latter type of limestone is often used as a building material, and can be mistakenly called marble when it is polished. Fossiliferous limestone is highly variable in color and fossil texture. The nummulitic variety of Nile Valley fossiliferous limestone was used to construct the Great Pyramid. The pyramid was capped with marble which was removed to rebuild mosques in Cairo after earthquakes. It is suggested by mathematician Stan Tenen that the original use of the Great Pyramid was to facilitate shapeshifting and time travel through sound activation of the limestone and marble. The state stone of Indiana and Tennessee is limestone.

LIMONITE: Limonite usually exhibits a rusty look and is sometimes cubic like pyrite (which it has replaced). More often a matrix for other minerals, it provides the electrical energy to boost their properties. Limonite is related to goethite, and is an affordable expression of goethite's abilities. It is grounding and centering, and assists with all blood and balance issues. *(1st chakra)*

LINARITE: Linarite is a hydrous lead copper sulfate which forms in a bright, two-tone blue coating. Linarite, energetically, will either impress the user with its properties, or the properties will not be apparent due to their subtlety. Linarite is highly energy specific. It is good for meditation enhancement and dream support. After major shifts in body chemistry or awareness, linarite is often useful to those who felt no attraction previously. It tends to ground energetic awareness on the psychic level. It deepens psychic processes especially for assisting investigative work, especially around homicides and abductions. Much like a bloodhound able to discern only molecules of scent in following a trail, linarite supports distinguishing highly subtle energetic clues in the environment. *(1st and 6th chakras)*

LINARITE and BROCHANTITE: This combination is so powerful that disorientation may occur after contact or use, so remaining quiet and calm afterwards provides the best and safest results.

LINGAM: A lingam, plural linga, is a stylized phallic symbol of procreation and the masculine cosmic principle. The word lingam means sign, and it is the common root of the Hindi words spade and phallus. Several types of rocks are utilized culturally for linga including jade, cinnabar and cryptocrystalline quartz. In India, the lingam stands in the middle of the yoni, or the center of the root (muladhara) chakra where it represents a light-lingam around which kundalini energy spirals up to the crown chakra and beyond. The lingam is considered the embodiment of Shiva, and sitting in its yoni allows life force to manifest. Lingam symbology in the East literally involves representing the central axis of life connecting Earth, the Middle World and Heaven. Many temples utilize a large central lingam surrounded in mandala fashion by eight smaller linga. Linga are used as standing stones for agricultural fertility and to mark mountain passes and peaks and road junctions as reminders of Great Mystery.

Any mineral can be used for linga; however, at one locality on the Narmada River in central India, red and brown chalcedony with iron oxide and goethite is worked through shaping and polishing to form a more balanced ritual object which combines masculine and feminine attributes, providing a unified vibration. These stones are called Tantric Linga, and are considered excellent for living spaces as they provide natural resonances to earth frequencies, grounding and balancing. As healing tools, linga respond to touch activation, and behave as all quartz does to piezoelectrically absorb and release energy. *(All chakras)*

LITHIUM: Lithium can be likened to the tooth fairy bringing a gift to the user when some part of the self which is no longer useful is shed. This is something like recognizing the rites of passage, and for the self to make comment on internal changes. It can be useful to hold lithium in meditation so one is induced to more clearly perceive the benefits of discarding what has become baggage. The name in spirit is "abracite," being especially useful as a stone to ward off calamity, as in the term "abracadabra." Most people will assume it is just a rock, just as they assume abracadabra is gibberish, therefore it can be worn and shown to others without hesitation.

Lithium has transcendental powers, making it possible for change and transformation to occur more rapidly, peacefully and equably. It is a power stone, enhancing clarity and the ability to stretch out into transformative space as defined by the individual before complete entrance into that space, so that prior information can be gained for preparatory purposes. It is unequaled as an agent of emotional balance. The element lithium is not indigenous to earth, and has come here through cosmic radiation only. It is very light, and it turns minerals pink, lavender or purple. It is the third element in the Periodic Table of the Elements, and it is literally the bridge between energy (the lavender ray of the life force) and matter into the nervous system which is the first body structure built when the cells of an embryo begin to differentiate. Without lithium, there would be no tolerating the energy blasts that the body must negotiate in order to be physical. The nightshade family of plants, or solanaceae, contain lithium in appreciable quantities. They are potatoes, tomatoes, tobacco and eggplant. It is no wonder why many shamans smoke, and why tobacco is a sacred plant to Native Americans. Lithium is located in the Spiritual Apothecary in the Auric Conditioners division. *(All chakras)*

LIZARD STONE: Lizard stone is a tuff of nickel oxide. It exhibits a yellow-green shade. It is good in the kitchen area for many general purposes. *(3rd and 4th chakras)*

LODESTONE (Magnetite): Lodestone is a double iron oxide often occurring as small, rusty octahedrons. It exhibits the quality of polarity or magnetism. Magnetite measurements taken sequentially across the Atlantic Ocean have shown that earth's magnetic field periodically shifts from the north pole to the south pole. Magnetite strengthens the auric field and helps rid the body of toxins. It promotes stability. By balancing polarities (yin and yang), it enhances overall healing and well-being as well as personal charisma. It increases personal power, and attracts people to the One. It is used to charge pyramids and crystal formations when placed at corners. This placement may more than double the grid power. *(1st chakra)*

MAGNESITE: Magnesite is a magnesium carbonate. It is white and opaque. Magnesite is excellent for combating genital diseases and for reassessing one's opinions or experiences of sexuality. It allows an individual to look truthfully at disconcerting or painful issues involving the first chakra, and promotes the conviction to change damaging or negative behaviors. *(1st chakra)*

MAGNESITE in or on QUARTZ: Magnesite can be found as a white inclusion, but it can also form as a coating on the outside of quartz, where the deposition of quartz did not occur after the magnesite. It draws impurities from the third eye. Place on third eye chakra for 15 minutes each week. Cleanse after use. It is useful for clearing past and present life trauma through visualization, hypnosis, energy work, etc. *(6th and 7th chakras)*

MAGNETITE: See Lodestone.

MALACHITE: Malachite is a hydrous copper carbonate mineral which is often deeply green. It forms in many styles from massive chunks good for carving, environmental and healing purposes to fine blades called acicular growths. This powerful healing stone can be placed directly on an affected area to absorb illness energies, whether they are physical or emotional problems. (Cleanse after use.) It moves energy from the higher planes, and grounds it into the green Earth. When used in meditation or focused concentration, it brings out the true reality of the subject to the person.

You will note that this stone is a team player, and commonly shares space and properties with azurite and chrysocolla. Malachite is a warm and accessible mineral with properties of importance to physical healing work especially in the upper chest. It cannot be overused, as it resonates deeply with human frequencies. It helps with heart, liver, lung, thymus, bone and spinal issues. This form of malachite is feminine. Expect it to work most efficiently in a cyclic pattern which becomes evident over a period of months, and then focus attention in healing applications to those periods when it is most strong for the individual. Without taking note of this cyclic pattern, one may tend to disregard malachite's usefulness as a healing tool. It is applicable to arthritis, bursitis, sciatica, scoliosis, broken bones, osteoporosis and rheumatism. Malachite is located in the Spiritual Apothecary under the Etheric Cleansers division. *(4th chakra)*

MANGANITE: Manganite is manganese oxide hydroxide. This velvety mineral has many cousins, but is itself quite simple. It promotes focus, simplicity, direction. It is centering and energizing through the heart chakra. It operates best on the heart chakra by simplifying decisions of the heart. *(4th chakra)*

MANGANOCALCITE: Manganocalcite is a manganese-rich calcite found in South America. It is pink and occasionally botryoidal like smithsonite. Also, it is found massive, allowing it to be tumbled and carved. Manganocalcite properties are closely aligned to anhydrite. Manganocalcite aids the liver, sexual impotence, the nervous system and lactation. It also helps with blood clotting, enzyme production, anxiety, stress and irritability. Manganocalcite has a female, receptive vibration and promotes connection to female spirit guides and female tone angels. It can be called the mother stone, as it transmits mother energies of giving life, nurturing and love. Manganocalcite also is a truly neutral mineral form of the Cosmic egg, therefore it is the embodiment of wholeness, oneness, bisexuality, the hermaphrodite, and the first cause of manifestation. It represents undifferentiated being, spiritual perfection, purity and the ability to discern flaws in others. Manganocalcite symbolizes the immaculate conception wherein the Virgin conceived by the Holy Spirit. It is best cleansed by moonlight and with sound. Manganocalcite resonates to the unicorn. *(4th chakra)*

MANGANOTANTALITE: Manganotantalite is a simple manganese tantalum oxide which forms a rich magenta, heavy tabular crystal. It occurs massive and can be brownish black. Called "The Harvest stone" metaphysically, manganotantalite emits the frequency of "all in its right time." It emits a frequency of gestation, expectation, and anticipation as well as letting go into completion, receiving the benefits of cultivation and intention, and knowing when the harvest is ready or ripe for the taking. It speaks to fecundity and maturity. Manganotantalite can be thought of as a depth finder, emitting signals to test depth and receiving back an image of hidden or unseen terrain. In human relations, this translates as the ability to recognize greatness or depth in others, and report back to them what is observed for their highest good. Manganotantalite emits a loving, gracious energy of bountiful welcome, setting a place for The Stranger, and making room for all. *(1st and 4th chakras)*

MARBLE: Marble is metamorphosed limestone, commonly exhibiting an opaque white color with patches of green, gray, brown, red or pink. It is a stone of the third ray. The different color variations will impact the body according to the normal color/chakra associations. Marble, whether viewed in architecture or hand held, communicates mobility within proportion. Residing within it is an inherent order to its apparent randomness of design. This mineral has a balancing quality like soft, but intricate, music which creates a tender link with the soul. The pink variety soothes hearts and grounds our love to the physical plane. Unequaled as a building or sculpture stone, marble carries the sound sensitive qualities of calcite. Wherever it is used in building material, it enhances psychic reception, connection to the other side, and shapeshifting. The state stone of Alabama is marble. *(Varies with color)*

MARCASITE: Marcasite is an iron sulfide. Marcasite was a popular stone in turn-of-the-century jewelry, as it provides a beautiful iridescent flash when faceted. Remarkably, sulfur supported home health remedies of body purification in the form of sulfur and molasses tonic. Marcasite still provides this cleansing vibration today. Marcasite cleanses the bloodstream, and it is a mineral which spring cleans the body. It supports the B vitamins to strengthen nerves and metabolism. It helps to build amino acids. Yet, it possesses a distinct aura of spirit vibration which makes it resonate to themes expressed in Richard Bach's book *Illusions*. Down-home Midwestern folks can fly, too, and they can experience a great fluidity in relationship to their higher selves, much like fields of wheat rippling in the wind. What could be so grounded and yet so open to the immeasurable pulse of consciousness upon wide plains of land? There is a great depth of knowing yet to emerge from the constancy of Midwest people, and their enlightenment shall be a blessing to all. For those who feel wings on their feet, marcasite is a grounding and yet celestially affirming stone friend. Marcasite speaks to the edge and transition awareness of people who feel they have been abducted by ETs. It also connects individuals to the larger consciousness that Earth is our home, indivisible, fragile, iron-hearted. It is difficult to comprehend what cannot be held in the palm of the hand for scrutiny, because we are at once the palm and held in the palm of life. *(Auric chakras)*

MARCASITE and GALENA with QUARTZ: This is a massive iron sulfide mixture which can be cut into various shapes. "Celestial soda pop." There's a song by that name which aptly describes this combination of minerals. "Don't leave home without it" is another good description. The point being that this combination is happening, it is highly energetic, and promotes fancy as well as imaginative insight. It lightens the stress of being overwhelmed by the Earth school and it provides connection to the higher self. *(All chakras)*

MARGARITE: Margarite is a calcium aluminum silicate classified in the mica group. It is a psychic intake mineral. It supports grounding during psychic work and clear receivership. *(1st and 6th chakras)*

MARIPOSITE: Mariposite (chromian phengite) is a woven, massive rock found in Mariposa, California, and is composed of calcite, chromium, manganese and occasionally sulfur. Mariposite is primarily green and white, with traces of black and sometimes yellow. The amazing aspect of mariposite is how it speaks to resiliency. The Spanish word for butterfly is *mariposa*, and the fine delicacy of this insect is also an aspect of mariposite. How does the short life of a butterfly become enduring and resilient to environmental forces? It chooses stages of lifeform where it can incorporate genetic response to stress in the environment, such as the worm or caterpillar state. It houses itself in a tough material such as a silk cocoon, and then it lives a life of brief, gentle, captivating beauty. In rocks, brief is usually exemplified in the aspect of shattering. Mariposite is a rock which experiences a shatter phase, and then a quick, intense conglomeration which creates the strength of character and durability exemplified by silk. Mariposite supports the ability of the individual to possess resilience of personality while going through transformation. Without it, the shatter phase of existence creates separation rather than reorganization and union with the principle of molecular adherence. It is a truth that some molecules desire bonding. The inability to bond is the recognition that surrounding surfaces or molecules are in trauma, and the original molecule can "absorb" trauma because it has no other option except complete and total separation. The risk in accepting separation is that it is an insurmountable barrier to unity. Mimicking the surrounding tissue until release from the situation is common, and leads to the ability to reprogram and adapt and thrive in new conditions. Mariposite embodies the essence of butterfly medicine. *(Etheric chakras)*

MARMATITE: Marmatite is a black ferruginous variety of sphalerite containing up to 20 percent iron. It provides grounding and connection to Earth energy. *(1st chakra)*

MERCURY: Mercury is a metallic element, liquid at room temperature, which is poisonous, because it can disable enzymes in the body. Another name for mercury is quicksilver. Mercury is an essential component of alchemy, and early on the name mercury became synonymous with the practice of alchemy. Mercury is a god of communication, trickster energy (as in coyote medicine), magic tricks, stealing, lying, and foolish behavior. The role of Mercury or coyote is to bring laughter and lightness to an overly serious estimation of self or a situation, much like the court jester. It's important not to assume that the jester is a fool or crazy, because underlying the mask is a wily, shrewd observer of the human condition and life. In Chinese alchemy, mercury was considered feminine, corresponded to the dragon, semen and blood, the kidneys and the water element. In India, mercury was a solar symbol concentrated underground. "Mercury knowledge" is another name for Yoga, the science of inner regeneration. Cinnabar is the perfect marriage of mercury (soul) and sulfur (spirit) alchemically, and to obtain mercurial energies, it is suggested to utilize mercury minerals rather than elemental mercury. Mercury as a native element is very rare, and cinnabar is its most important ore. Mercurichrome is an antibacterial tincture of mercury. *(1st through 6th chakras)*

MESOLITE: Mesolite is a hydrated aluminosilicate of sodium and calcium listed in the zeolite group. It is a white acicular zeolite similar to natrolite in appearance and properties. It forms chemically midway in a series between scolecite and natrolite. Mesolite is usually found in delicate specimens not given to handling, so its properties are best obtained through proximity in the environment. Mesolite applies well to intimate, one-on-one contacts in a healing setting such as individual counseling, massage, hypnotherapy, energy work, etc. It opens both the fifth and seventh chakras. It is a good choice for sleeping areas as maximum benefit is obtained from abundant daily exposure. It stimulates a form of internal sight awareness which is useful to seeing-eye dogs and dogs which are able to detect the immanence of seizures. It may stimulate foreknowledge for those suffering from seizures, so they can get themselves to a safe place. See Thomsonite. *(5th and 7th chakras)*

METALS: A great portion of the elements are metallic or semi-metallic. The ancients did not have the benefit of present chemical knowledge, but many traditions recognized essential aspects of metal behavior and characteristics through observation. All metals are energy grounders, but some are capable of simultaneous ejection of awareness to out-of-body states, in particular marcasite. In many cultures, metals were and are considered impure, probably because of their mining associations to the underworld. Blacksmiths could either be rejected from society or thought to be agents of transmutation. The alchemists' Great Work was essentially metal-based and metal-oriented. Whether alchemists literally hoped to transmute lead into gold or used the process as an allegorical and symbolic representation of human spiritual transformation seems best answered by assuming they meant both. In the Masonic tradition, laying aside all metals in certain initiation rites was an indication of laying aside all material goods and attachments in order to regain innocence. Other practices today exemplify this laying down of metal, such as airport security check points and metal detectors on school campuses. The need to lay down one's metal to shun violence is as valid and necessary today as detachment from material goods has been historically.

It is interesting to note that the planets are associated with metals, and the farther the planet is from earth, the metal is considered lower or less exalted. The following are the associations: Sun-gold, Moon-silver, Mercury-mercury, Venus-copper, Mars-iron, Jupiter-tin, and Saturn-lead. Also see the listing for each individual metal. Metals are the most economically valuable minerals, and many rare species of minerals are sacrificed to mining expediency where normally at least 100,000 tons of ore must be processed each day to meet economic quotas. Often the finest mineral specimens are found in exactly these mines. It was more common for miners to preserve specimens in the past, but modern extraction requirements usually force companies to fire miners who attempt to save specimens and slow down work schedules. Unfortunately, mineral specimens are endangered and lost forever like plant and animal species. Mineral characteristics are unique in each locality around the globe where they are found. Miners who risk their jobs to preserve mineral specimens are actually liberating minerals from complete extinction. Hopefully, as mineral collection increases, mine owners and operators will see that specimen sales are an excellent reason to halt mining operations long enough to preserve specimens. *(Usually 1st chakra)*

METEORITE: Usually originating from the asteroid belt and sometimes exploded planetary bodies, meteorites are typically metallic, because they survive entering Earth's atmosphere. They are coated black, brown or gray. Often a crystallized pattern is found inside a cut meteorite, and traces of rare irons and olivine minerals have been found in pockets of meteorites. The crystallized pattern results from cooling at an incomprehensibly slow rate. Meteorites are said to cool one degree every one million years. This allows the metals to express their crystalline characteristics which metals on earth cannot achieve with the same distinction. Meteorites are the survivalists of outer space, having traversed essential vacuums of nothingness to bring metallic coherence to earth. This highly evolved quality rubs off on humans by tending to help our consciousness evolve. Meditation with a piece of meteorite helps a person connect with other planet past lives and present extraterrestrial associations. It enhances telepathy, ability to cognize complex orders of universal reality, and it switches centers in the brain from reptilian dominance under stress to higher order noetic potentials. Meteorite increases order and decreases sensations of chaos and losing touch with reality. Meteorite increases pattern recognition and the ability to think globally, if not universally. It is highly attractive to logical and organized thinkers who are curious. These people are usually stubborn or well able to hold patterns for others, and they have brought in mutant genes to shift human development. If they do not bear children, their ideas will change the world, and it is to the benefit of the world that these individuals focus on ideas now rather than wait for their children to effect planetary change. Time is of the essence, and meteorite has waited a long time to chime in and activate consciousness shift. *(All chakras)*

MICA: The mica group is a complex association of layer silicates, most notably containing vacancies in their chemical structure, and usually dominated by potassium. Other elements which can be included are calcium, aluminum, barium, iron, magnesium, lithium, vanadium, chromium, zinc, manganese, sodium, and cesium. They exhibit perfect basal cleavage. Biotite, lepidolite and muscovite micas all possess unique assemblages of aluminum in their molecular structures. Mica sheets were used in turn-of-the-century phonographs and as window shields in wood stoves. All micas increase the ability to think in analytical, structural patterns conducive to physics, mathematics and engineering, thereby enhancing the ability to formulate sequential mental processes similar to mechanical processes. Micas tend to link abstract and logical pathways in the brain allowing humans to conceptualize how to use machines and technology in new ways. Fractal mathematics is an excellent example of how micas refer back to the source in order to create linkages. For the Hindus, mica combined with mercury provides immortality.

Biotite - See Biotite.

Celadonite - See Celadonite.

Chlorite - This green inclusion is common in quartz and as a coating on geodes.

Golden - These are the more common varieties, and yet their high metal appearance provides windows to the higher realms. Mica, usually found mixed with other minerals, is an ambassador (in the mineral kingdom) for the universality of things, inanimate objects, and machines. Golden mica melds energies and makes linkages among differences. It is a peacemaker. Mica allows the whole to be greater than the sum of parts. *(Upper chakras)*

Lavender - See Lepidolite.

Silver - See Muscovite.

Pink - Pink mica is an excellent heart chakra equalizer. The window rock qualities of mica in the pink ray are attuned to emotional revelation. When individuals experience confusion or hesitancy, meditation with pink mica often allows internal illumination so that the disrupting or causal factors may be identified. Pink mica does not possess a large field of influence in small quantities, so placing on the skin is suggested. *(4th chakra)*

MICROCLINE: Microcline is a potassium aluminum silicate classified in the feldspar group. Microcline is an energizer which also facilitates all creative and healing pursuits involving the hands, whether this is cooking, massage and energy work, art, writing, gardening, etc. Microcline increases the flow of energy from above the head and shoulders and down into the arms, alleviating meridian burnout and arthritic symptoms. *(7th chakra)*

MILLERITE: Millerite is nickel sulfide, and is called "capillary pyrite." It forms in green, hair-like whiskers. Because of its fragility and minimal mass, it often is found in vugs which protect it. It is not typically amenable to physical healing work.

MIMETITE: Mimetite is a lead arsenate chloride. It is usually yellow. Mimetite alone is good for supporting self-realization and recognition of life purpose. Combined with lepidolite, mimetite is useful in reestablishing safe bonds to reality for individuals who have repressed painful memories. Mimetite aids repair of synapse function. *(3rd chakra)*

MOLDAVITE TEKTITE: Moldavite is a green glass product of soil vitrification from a meteorite strike. It is the only gem quality tektite found worldwide, and originates in Moldavia, Czechoslovakia. This is a seer's stone to be used by the initiated and the adept. It will cause electrical impulse/frequency alterations on a temporary or semi-permanent basis, depending on the frequency of use. The green ray applied to the third eye automatically provides healing knowledge to be available to the user, and promotes healing of the brain/nervous system network, especially by consciously drawing white light through the chakras while having it on the third eye. It will heal coma induced by intense body trauma. It should not be used by people in shock as the blood chemistry will become confused by the application while the person is in this state. It will aid all disorders stemming from improperly firing brain connections. It can be used to create bridges between synapses which no longer function, but it is not a permanent remedy. Synapse dysfunction healing would require the knowledge of a skilled practitioner, as the laser-like healing beams can also create other damage if not properly concentrated or positioned. Moldavite is an Auric Sealer in the Spiritual Apothecary. *(6th, and high-range, upper)*

MOOKAITE: This Australian jasper-like stone reflects the Outback's rustic pastel shades. It is a transpiration supportive stone which assists in the transference of spirit essence into physical. It can apply to the moment of fertilization of an egg, as well as the moment of spirit entrance into the fetus, therefore it is beneficial for gestating women. It is a rainmaker stone which aids the calling of thunder clouds and rain. It assists gardens, especially in dry climates. It can be used to overcome dehydration. *(1st chakra)*

MOONSTONE: Moonstone, also known as microperthite, is a gem variety of orthoclase which is a potassium aluminum silicate. Moonstone blends and balances emotions. Besides the normal heart placement in healing layouts, it can be used on the will center to link to higher consciousness with protection. Moonstone specifically synchronizes moon-based cycles (menstrual and biorhythm), and aids women in all menstrual transitions from PMS/monthly, to menarche and menopause. It is an excellent stone for the moon lodge and altar, and an excellent gift to give a young woman after her first menstruation as a symbol of the time of making life wishes. It is an excellent stone to balance female fertility issues in humans, animals and the land. Moonstone can lessen the impact of an overly masculine personality. It is a mineral which supports fine-tuning emotional and intuitive awareness rather than focusing on release of emotional or psychological trauma. Because it resonates to the detached energy of lunar essence, moonstone resonates to observation, receptivity and influence through vibratory effect rather than force or action. It offers the essence of the Tao, nonaction, and holds center in the constant surge and ebb of energy in the flow of life. As this energy becomes harmonic within the being, moonstone allows the feminine aspect of reason and thought to become known and understood through reflection. Moonstone exhibits a stoic form of rationality, the ability to withstand the demands of a changing natural world, because one always has withstood. Moonstone enhances theoretical and conceptual thinking as well as thinking globally and inclusively.

Moonstone brings release from excess, and holds both passion and pain at arms length. It is a stone which honors the whole, the Middle Way, the path of constant dissipation of structure. Energetically, moonstone provides a structure which always shifts, changes, evolves, mutates and absorbs. It is related harmonically to owl, the element water, and the astrological sign of Cancer. Moonstone harmonizes well with people who care for the land through farming, gardening, landscaping, feng shui and environmental work. Moonstone is a mineral like other forms of orthoclase which guides people through the dark, whether this dark is a time of questioning and seeking or deliberate vision quest. Utilizing several moonstones on an altar to mark the phases of the moon for ceremonial purposes is an excellent way to physically mirror the increase/decrease cycle of the moon. Moonstone is an excellent choice for deepening connection to the unconscious and dream state. Moonstone is the state gem of Florida. Moonstone is the June birthstone. *(4th chakra)*

Clear/White - Clear moonstone is a total body integrator, bringing together the etheric, physical and auric levels. It is highly valued in the spirit realm as a mineral which allows physical beings to be perceived accurately in totality, meaning that people who work constantly in the spirit realm maintain clean surfaces where spirits can access them and readily understand intent, vibrancy level, skill level and transmission clarity. Clear moonstone is an absolute must for all energy workers, healers, psychics and dreamers. This form of moonstone resonates to lizard. Clear and white moonstones are Etheric Sealers in the Spiritual Apothecary. *(All chakras)*

Gray/Blue - Gray moonstone and catlinite, or pipestone, are a powerful mineral combination resonating to wolf clan energies. This tone embodies the energy of educators and teachers, bringing a wealth of experience and knowledge in just the areas necessary for personal growth. Moonstone which is definitely gray is a rain energy mineral and is useful in calling rain spirits.

Green - The gentle ebb and flow of moonstone is given physical sensitivity in this color tone, which counteracts all diseases involving hardening of tissue such as multiple sclerosis and scleroderma. Green moonstone enhances flexibility and equilibrium. It is an important emotional balancing stone which provides soothing vibrations. *(4th chakra)*

Peach - For "people who give too much" with sore and abused hearts, peach moonstone is a caring friend. It soothes and strengthens overtaxed tissues of the heart, releasing accumulated issues, sufferings and toxins absorbed for other people. It provides emotional relief, endurance, and greater clarity about boundaries of giving too much. For those who don't give enough and have been called selfish, peach moonstone strengthens heart resonance to plants, animals and humans, increasing charisma, compassion and empathic awareness. It enables individuals to offer themselves with increased unconditionality, sense of service and respect for the needs of others. Peach moonstone represents the dignity of the lion-hearted. *(4th chakra)*

Rainbow - Rainbow moonstone acts similarly as spectrolite, chalcopyrite and labradorite, but its unique aspect is the ability to address auric disturbances and imbalances in chakras above the head. Rainbow moonstone works primarily outside the body on the spirit level. Because it supplies stronger connection to spirit realms, rainbow moonstone can facilitate a physical reaction where the endocrine system is rebuilt and strengthened. Spiritual malaise or issues often cause endocrine upsets. Rainbow moonstone is an Auric Conditioner in the Spiritual Apothecary. *(Upper and auric chakras)*

MORGANITE: Morganite is the form of beryl which is light pink. It is named after J.P. Morgan, who funded the California exploration where morganite was first discovered and identified. Morganite applies to the heart center. It provides strengthening for the muscles and nerves of the lungs and throat area. Its wet pink look provides an opener for releasing pent up emotions when visualization work is done, possibly causing tears of empathy. Morganite embodies the spirit of serpent as teacher, the harbinger of knowledge and wisdom. It enhances willingness to forgive, release and move on. It is a member of the Spiritual Apothecary in the Healing division, and among the beryls is the representative of healing power.

Morganite resonates to air energy, which is typically identified as a masculine element; however, morganite literally carries the breath of life principle and shares with emerald dominion over the heart chakra. Morganite brings light to the seat of the self as well as the energy of the Godhead, which is normally viewed as the originating and neutral source of the more masculine sky divinities. Morganite is the home of the gods on earth, therefore morganite literally connects to heaven. Heaven, as universal spirit, is the place where the human soul may rest when it has come to perfection. Achieving this often involves being split and opened by spirit in a lightning burst of revelation and enlightenment. Morganite infuses the subtle and physical bodies with unconditional love, because it is the cracked halves of the primordial egg of Heaven and Earth split open to reveal Love within. Morganite resonates to dove energies of love, peace, the imperishable spirit, and Aphrodite's bird. Morganite is an excellent companion stone to petalite, danburite, tourmaline and phenakite. *(4th to upper chakras)*

MORRISONITE: Morrisonite is a multi-colored jasper-like stone. When placed over the third eye in a prone position, morrisonite relieves ear and sinus blockages. It also embodies the magical and healing qualities of the heron, especially the Great Blue Heron. In ancient times, the people told the story of Hitatsa, a venerable medicine woman renowned for her healing prowess. As she neared the end of her journey in that life, she contemplated many long months on what great gifts she could leave for the Creator in thanks for such a transcendent life. One afternoon, as she stood on a high bluff overlooking the ocean, watching storm clouds recede to reveal the sun and an amphitheater of rainbows over the water, Hitatsa saw in a vision a flock of Great Blue Herons fly over her head and into the trees behind her. She had never seen such magnificent beauty in flight as was revealed to her at that moment. She asked to be released into spirit if the flock could be transformed into physical reality. Her wish was granted, and the ancients believed for millennia that the feathers of the Great Blue Heron, only when found and not taken from the animal, were superior healing medicine. As so often happens between the kingdoms, Hitatsa's honor was transformed into mineral form also. *(5th and 6th chakras)*

MOTTRAMITE: Mottramite is a hydrous lead copper vanadate which exhibits a black botryoidal growth. It is classified with descloizite, where zinc replaces the copper, in the descloizite group. Mottramite is related to vanadinite's ability to manifest on the physical plane. Mottramite strengthens connection to higher self, spirit guides and soul group. It brings higher spiritual ideals to ground in this life, thereby stimulating the ability to manifest spiritual goals, values, visions and gifts as well as life force energy. Mottramite enables the body to serve as the temple of the soul, and it assists the body to overcome physical barriers, especially through deepening focus and intention. Mottramite reacts with the body to reduce fever. It should not be used by people with low blood pressure or poor circulation, because it tends to depress heart rate slightly. It can be useful to curb high blood pressure and to understand the body's biofeedback. *(1st chakra)*

MOZARKITE: Mozarkite is the Missouri state stone, and is a grainy, pastel mix. Mozarkite is a stone for elementals or devas, and is useful for working with energies of new homes. The mineral promotes attunement to energy lines of the property and to all the devic forms living there. If a person travels a great deal, it facilitates quick alignment with a place to enhance sleep, overall relaxation and digestion. *(1st and 2nd chakras)*

MTOROLITE: A silky, velvet textured green chrome chalcedony from Africa, mtorolite is a natural support stone for all practices involving homeopathy, plant essences, aromatherapy and plant cultivation. It is especially useful as a stone which helps to "weather the storm," providing a sense of lucidity in crisis and the strength to endure. It supports the heart chakra. *(4th chakra)*

MUSCOVITE (Lavender Mica): Here is a story: Once, while sailing on thermals, a mountain eagle saw a lovely shiny object far below, but as he swooped down to inspect it, it lost its luster and he lost track of it. He spiraled up again, and was watching the ground for movement when he spied that shininess again. This time, he found several objects to home in on so he wouldn't lose the beautiful object, but once again he failed to see it when he swooped low. He left off searching for the object, but kept it in this thoughts all the night long. At dawn's earliest light, he was up in the sky again, searching. He knew now that he would need help finding the object, but he didn't know whose aid to enlist. The crows and other birds always drove him away. The rabbits, shrews, and mice were his prey. The coyotes, wolves and large cats were his equals, but too busy with their own work. As he thought through the list of possible friends, he realized there were none, and he felt sad. At that moment, the shiny object caught his eye once more. The eagle settled on a high limb and watched the object all day. Occasionally it wouldn't shine, but as the sun found it again, it would bloom forth in brilliant lavender color. By and by, through the forest came a child wandering aimlessly. It came into the clearing and suddenly looked frightened. The eagle watched, and came to realize the child was lost. "Ho there," he called to the child. The child searched the sky for the voice, and saw only the beautiful eagle on a bare fir tree branch. "Ho," the eagle lifted his wings in greeting. The child was tired and too frightened to speak. The eagle spoke lovingly to the child, and learned that he had wandered too far to return home by sunset. They agreed the child would sleep at the base of the tree, and the eagle would watch through the night. The eagle saw many things — bats and owls and bobcats — but not one came close to the tree. In the morning, the child found berries, and drank from a stream. Just as he and the eagle were setting off to find his home, which the eagle could see on the horizon, the child realized he could not thank the eagle enough for being so kind and being his friend. Just then, the eagle realized his friend had come to him, and he asked the boy to retrieve the lavender rock. The boy did so, and instantaneously, the boy and the eagle were in the air, flying toward his home…

This stone aids in raising the vibratory pattern of the person so that the higher self can be contacted. It works in harmony with the crown chakra. In combination with stones which work specifically with each chakra, muscovite amplifies and aids the vibration thereof. Muscovite also aids in the contact of spirit guides. All of the above is a lovely way of making clear an important facet of muscovite's molecular structure which is that it has a vacancy, as several forms of mica do. This vacancy allows oneself to get out of the way in order to receive guidance and support. It helps to highlight how self-sabotage is operating, and how to initiate behaviors which will result in resolution and solution of problems or challenges. It aids in recognizing friends and placing trust in them. Muscovite is listed in the Spiritual Apothecary in the Auric Sealer division. *(6th and 7th chakras)*

NAMBULITE: Nambulite is a lithium sodium manganese silicate hydroxide. It carries an energy of tender cleansing which makes it quite good for detoxifying soft tissues, especially the skin and organs. The liver and lungs respond readily to healing applications of nambulite, particularly for cirrhosis, pneumonia and cancer of either organ. While providing a strong physical benefit, it releases emotional and mental issues which weaken the body in the soft tissues, and helps to ward off return to the old, familiar emotional or mental patterns. Nambulite is a higher order expression of rhodonite. Nambulite's focus on the liver rather than the heart (rhodonite) highlights the importance of the liver as an organ of emotional awareness. In the past, the liver was considered the seat of the emotions. It is the only organ which can regenerate itself if given the opportunity. Nambulite assists this healing. *(4th chakra)*

NATROLITE: Natrolite is a hydrated sodium aluminum silicate, and is classified as a member of the zeolite group. It is sometimes known as "white cat's eye" in its more massive form. It has fuzzy spires that look like fine needles on a pin cushion. The white color and etheric nature of its appearance causes a subtle crown chakra opening. *(7th chakra)*

NEPTUNITE: Neptunite is a potassium sodium lithium iron manganese titanium silicate. Expressing itself in a small, black, square crystal, neptunite is often found associated with benitoite. Neptunite harmonizes well within a network of other stones, and acts as a transmitter. It is attuned to Neptune-born individuals. Neptunite is a "course corrector," allowing group dynamics to be maintained without a disciplined structure. Its essence allows individuals to orient to the broader community vision, and promotes voluntary contribution. The sense of neptunite is to recognize and accept one's obligations, which can be embraced with a free and open willingness. One can recognize the inherent demand of the activity, but it is raised to a vibration of service and contribution rather than enslavement and demand. It promotes circular sharing of power within a group rather than hierarchical structure. Where cooperation is required, neptunite reaches out to create a community allegiance, and allows individuals to highlight the aspects of group intuition. This can lead to a prophetic ability in describing the destiny of a culture or group, and allows the individual to play an important role. For those requiring drugs to maintain interest in obligations, and therefore one's role in life, neptunite can be balancing and harmonizing. It is useful for overcoming addictions. It is named for Neptune, the Roman god of the sea. Neptunite increases kundalini flow when placed near the root chakra. It is recommended to begin kundalini practice with other minerals first. See kundalini in index. *(1st through 7th chakras)*

NIPIGON STONE: "Shepherd stone." This stone is porous and brick red with yellow spots. For those who have responsible, yet monotonous and boring jobs, this stone strengthens intention to remain present and alert. Even as the individual may catnap or daydream or watch a distant sky, there is an underlying alertness at all times, which allows the individual to monitor what occurs in the vicinity. It is also good for people who want to blend into the background as they observe others, such as researchers or detectives. It provides natural camouflage. Nipigon stone embodies fox medicine. *(6th chakra)*

NUNDERITE: Nunderite, a mauve and green-spotted granitic stone from Australia, is another natural representative of the pink-and-green combination stones. It is a gemmier expression of turtle back stone energy. Where turtle back is an expression of water and earth elements, Nunderite is an expression of earth and fire, a fusing of molten elements. Coming from the continent where the Heart of the Mother resides (Ayers Rock or Uluru), this stone is capable of providing the same Logos energy, however rather than the sacred moment of fertilization as in turtle back stone, this stone embodies the sacred moment of transformation of energy or transmogrification. While fertilization is one aspect of creation, the destructive capacity of molten earth becomes her way of creating herself anew. *(4th chakra)*

OAXACA "DIAMOND": Oaxaca diamonds are quartz crystals, shaped similar to Herkimer diamonds, found in clay in Oaxaca, Mexico. They are the masculine counterpart to feminine energy Herkimer diamonds. They provide many of the same qualities as Herkimers. They are referred to as "God's Storehouse" metaphysically due to their ability to receive programming. *(7th chakra)*

OBSIDIAN: Obsidian is volcanic glass, and forms as a result of magma reaching the surface of earth where it cools. The method of cooling affects the character of obsidian. Most forms are found in desert regions. All forms of obsidian are valuable as protection stones, especially for sensitive, empathic or psychic people. It grounds higher energies through the base chakra, thereby balancing the digestive and intestinal systems. It is a good pocket stone. All forms of obsidian are listed as Healers and Auric Sealers in the Spiritual Apothecary. The following obsidian varieties also have additional attributes:

(NOTE: All obsidian types are 1st chakra with additional influences noted.)

Apache Tear (Transparent Obsidian) - Provides psychic protection and grounding as symbolized by the light shining through the darkness. Draws light to the base chakra. A psychic shield. *(1st chakra)*

Black - Black obsidian is the Wizard Stone. It not only has a sphere of influence which holds back negativity, but in a large mass is a "door opener" both within an individual and in the psychic sense. It exists simultaneously in more than one dimension. Yet with all its power, it is a benevolent stone. It cannot be misused, for only positive vibrations can work harmoniously with its purpose. In the egg shape it draws upon itself infinitely. Purity of spirit is reflected by obsidian. Negativity cannot coexist with it, therefore its presence aids the positive vibratory force of those entities and those material surroundings within its force field. It can also draw healing to the asker. This stone is highly recommended for all psychics. *(1st and Auric chakras)*

Fire - A good stone for grounding kundalini energy which tends to rise too swiftly or powerfully. Once a practitioner is accustomed to this mineral, it tends to enhance awareness of the body's energy pathways enabling body workers and physical healers to locate blockages or weaknesses, etc. Fire obsidian is the highest union of igneous attributes. The glass quality combined with the fire aspects allows the mineral to bring clarity with often startling visions or intense energy. *(1st and 2nd chakras)*

Golden Sheen - This variety of obsidian combines grounding and protection traits with upper chakra golden hues to provide an integrating and invigorating effect when worn or applied to the base chakra. It brings the golden ray into the Earth energy area (1st chakra), and activates the will on the way. *(1st, 3rd, and upper chakras)*

Green - Green obsidian can often be colored a drab, olive, Army green. As a companion stone to other minerals, green obsidian provides energy to other crystals and minerals, and is a friend to them. It can be used to heal the trauma of dropped and chipped crystals.

Mahogany - Mahogany obsidian has brown, wood-like streaks in a black base. It inhibits the development of negative energies in the everyday, mundane environment. It is a worker stone, in the way worker bees keep the hive clean. Obsidians generally don't need cleansing attention, but this form will benefit from regular attention due to the amount of work it does.

Pink Snowflake - Pink obsidian exhibits pink bursts of ash in a black base. It is a crown, heart and base chakra balancer, integrator, harmonizer, and promoter. It is an excellent pocket, pouch or jewelry stone for people who lack cardinality (Aries, first house, Cancer, fourth house, Libra, seventh house, Capricorn, tenth house) placements in their horoscope. It is good for maintaining balance between physical, emotional, mental and spiritual aspects of self. It heightens physical sensitivity to and awareness of energies being transmitted by others, as well as helps to modulate the empathic effects of receiving psychic flashes. See snowflake obsidian. *(1st, 4th and 7th chakras)*

Rainbow - Rainbow obsidian shows layers of growth through beautiful, iridescent stripes running through a black/dark gray base. All the spectrum of light and color seeps out of this dark, glassy rock when broken or cut at the proper angle. Its full spectrum influence is a visual reminder of the wholeness of life, and speaks for itself on a subconscious and conscious level at once. *(All chakras)*

Root Beer - Root beer obsidian is rare. When held up to the light, it is transparent and colored just like root beer or brown bottle glass. When combined with aquamarine, emerald and heliodor, root beer obsidian completes a profound healing matrix. Root beer obsidian empowers and enhances the other three incredibly. Shaping it in cubes or pyramids optimizes its effect for healing work. For skilled practitioners, tubular wands are excellent tools for placing in the ears to correct hearing problems associated with high decibel damage. This crystal can be ground, and creates an excellent sacred sand. The refractory capability of the shards is beautiful for the visual impact of making sacred shapes. Shapes can be crafted and worn as a primary protection amulet for emergency workers. The healing strength of root beer obsidian revitalizes energy lost by accident victims. It infuses humans with a friendly, life-giving essence which feels extremely precious. It is powerful as a gift, because bringing receiver's healing abilities are blessed, celebrated and acknowledged by the transaction. Each small piece and shard is a treasure. This stone cannot be misused. It does not reveal its power to those out of integrity. *(All chakras)*

Sheen - When cut into balls or eggs, it typically forms concentric silver circles (eyes) which can be the focal point of meditation to tune to other planes.

Snowflake - Literally, snowflake acts as an intermediary rock where all colors combine to create white embodied in the black protection of obsidian. It is a good stone to use in times of change for a balanced attitude. It can be used in meditation, teleportation, astral travel, psychic connection, reconnection with soul group, planet or galaxy.

Called the "Book of Laws Stone," snowflake obsidian is the guardian of the texts of Great Spirit and ancient civilizations. It aids psychic hearing or clairaudience. The power of this stone derives from its multilevel psychic impact and characteristics. It integrates distant, past and future parts of self for an immediate embodiment of the total self. Snowflake obsidian stimulates and promotes transformation. It resonates to raccoon energy. *(All chakras)*

OBSIDIANITE: A category of manmade or remelted natural volcanic glasses. Effect varies with the color.

Emerald Green (Philippines) - This vibration of green is both responsive to healing energy and supportive of transmission of healing energy, therefore it could activate interesting resonance when used by one person on another. Used individually, it increases spirit guide instillation of healing energy. It can be used to great benefit to help spirits not yet disengaged from the Earth vibration. *(4th and Etheric chakras)*

Light Blue - This vibration of blue is high quality and intense. It often will harmonize with an individual more strongly than celestite. For people who communicate well, it enhances listening abilities. For people needing help in communication, it activates the ability to hear oneself speaking distinctly and persuasively. It aids in cellular retention of affirmative healing statements. *(5th chakra)*

OCEAN PICTURE STONE: Ocean picture stone is a copper silicate with opaque tendencies suggesting its classification is a jasper rather than an agate. This stone often resembles ocean waves crashing upon a rocky shore, as it bears sky blue to greenish tones mixed with white and brown matrix. It originates in one location in British Columbia, and is becoming treasured due to its rarity. It is an excellent soother, balancer, and harmonizer. It unleashes pent up tensions and anxieties in a positive manner, and provides a sense of scope or horizon to issues which cannot be resolved or seem too large to resolve. Ocean picture stone empowers sense of self and what can be realistically achieved. It increases electromagnetic sensitivity to earthquakes, energetic healing and the presence of spirits. It moves kundalini energy up and down the spine. Ocean picture stone embodies all the properties of ajoite in a more affordable form. It is found in the region of the Native American Nez Perce. It is important to note that Chief Joseph, of the Nez Perce, was the first to encourage his people to "fight no more forever" in modern times. In this light, ocean picture stone represents the sovereign who is thoughtful and methodical in action to avoid inciting his people to hasty displays of aggression. Ocean picture stone holds an energy of bringing all the margins and dispossessed into the whole for increased survival ability as well as community solidarity. Ocean picture stone is an Etheric Opener in the Spiritual Apothecary. *(1st, 4th, 5th, 7th and etheric chakras)*

Iron Oxide/Biotite Mica Included - When included with iron oxide and biotite mica, ocean picture rock brings etheric abilities into the physical, thereby deepening and strengthening the ability to resolve conflict and hold center. It draws power from the depth of being to be expressed through the throat chakra. Ocean picture stone resonates to elephant energy. See Turtle Back Stone. *(1st through etheric chakras)*

OKENITE: Okenite is a hydrated calcium silicate that looks like white "puff balls." It is so fragile that it is not typically useful metaphysically; however, its white nature impacts the seventh chakra. Specimens can be placed on a head board of a bed for regular exposure. It provides a marker or beacon for the wandering soul to return to during sleep. *(7th chakra)*

OLIVINE: Olivine is a magnesium iron silicate, which is a loose term for two minerals that form in a series with each other, forsterite and fayalite. Most olivine and even gem peridot is actually forsterite, which is a magnesium silicate. Many minerals result from the geologic alteration of olivine, the most common being serpentine. Olivine is an excellent purifier, especially when used after applications of bloodstone, citrine or smoky quartz. Olivine behaves more like an energetic fine tooth comb as opposed to a rake. It picks up small impurity fragments and attaches them to vitamin C and E for removal from the body. All fasts and colon cleanses followed by olivine essence and wearing of olivine for a week will produce noticeable results. Olivine tends to tax the system of minerals, so upping the intake of minerals during the "olivine week" is beneficial. *(All chakras)*

ONYX: Onyx is a variety of agate, which is a form of microcrystalline quartz classified under the heading of chalcedony quartz. It can be found in solid shades of color, but most frequently it tends to band in distinct layers. The pure, black variety is usually dyed. It is often used for carved products commercially, and is a silent infiltrator in many homes, bringing spiritual expansion qualities of which many consumers are not aware. It is worn too frequently or placed in the home by people who are not aware of what it can do for the experienced practitioner. It provides excellent resonance frequencies for meditation and dream emphasis. It is recommended to be used with apache tear (or other stones) for protection unless the practitioner is adept. It energizes body systems.

Black - Black onyx relates to the first chakra. A strong connection was made by early humans between what was eaten and how it later exited the body. Physically, it seemed, all things come from above and exit below. It was difficult to understand the connection between intercourse and the later arrival of children. It seemed woman must have swallowed something to make this phenomenon occur. The vaginal opening is unique, and shows that the most sacred process, bearing life, occurs opposite all the other beginning points for openings in the body. When humans realized the connection between the vagina and birth, the proximity to the elimination orifices caused a big stir. Menses must be bad and dangerous too! Interestingly, some cultures viewed the excrement of the gods to be sacred, and rituals built up around those beliefs. Black onyx is the stone of mystery surrounding the vagina, womb and ovaries. It represents the hanged woman at the mercy of divine will. It bears the depth of putting one's life at risk to create another. This is the Madonna or maiden aspect of the triune goddess. *(1st chakra)*

Green - Green onyx takes the ingredients supplied by the lungs and nutrients supplied by the digestive system, and pumps them through the body. It is the conductor of the heart bearing witness to new life in the womb and bringing life a new to the self. It is the premier generator of heart intelligence, allowing the heart to adapt to changes in cycles without skipping a beat. *(4th chakra)*

Honey - Honey onyx bears a light peach tone. It works on the second, third and fourth chakras to promote deep breathing for many uses, especially childbirth and pranayama. The mystery of breathing is as profound as the mystery spoken of for black onyx. Air enters the lungs to be integrated into the body, and carbon dioxide, a bi-product of respiration, is released. The nostrils represent the vagina of the lungs which can breath new life into others, projects, and self. The elements for life giving processes must enter rhythmically without significant pause to avoid brain damage. This constant cycle of life and death is played out continually in all manner of organic life, and honey onyx bears witness on the mineral level that constancy is a unified state of awareness. This is the crone phase of the triune goddess, and is expressed through stories from the heart, toning from the diaphragm, and knowing when to keep silent. *(2nd, 3rd and 4th chakras)*

White - White onyx is the holder of sacred space which processes what is seen and heard. It is the seed of greatness, the germ of consciousness. It is released through attention and devotion. It is the quest for the meaning and mystery of life, and the effort to restore the sacred language of life. It is the virgin phase of the triune goddess and expresses the young woman's responsibility to hold purity as the Holy Grail. It is knowing that humans, and in particular, pre-sexual youth, embody the chalice as the holder of the life-blood, and it is in losing one's virginity that innocence and the grail are lost. White onyx embodies the archetypal eucharist in which sacrifice is ingested in order to be made pure again as well as porcupine energy. *(7th chakra)*

OPAL: Opal is a hydrated silica which forms in microspheroids. Light bouncing off the spheroids causes opal's spectrum of color flash. When water dehydrates from opal, it can produce forms of chalcedony, such as agate, jasper, onyx, chrysoprase, bloodstone or flint. Opal is placed in the Etheric Conditioners category of the Spiritual Apothecary, because it harmonizes the emotional, mental, psychic, and supraconscious levels of awareness. In its red/orange (fire), black, jelly blue and white/yellow forms, its impact is accented for different body areas, acting as a tonic and regenerator. Opal is known to emphasize prevalent tendencies of a person, positive or negative. It is a powerful aid to clairvoyants. Common opal is useful to use in soul retrieval and entity removal work where the opal ritually receives the negative energy and is either burned, buried or crushed afterwards. Opal is listed as an Etheric Opener in the Spiritual Apothecary. Virgin Valley opal is the state gem of Nevada. Opal is the October birthstone. *(4th plus)*

Black (Australia)- "The Enchantment Stone." Black opal has been prized as a stone of immense magical power, adding to the superstitious lore of opal in general. It is the mineral precursor to raven and crow energies. It enlarges the sending scope of a long distance healer and increases discernment ability regarding the correctness of providing healing energy to another. A "wyrd" stone, black opal conjures up the fates for use by soothsayers, diviners and fortune tellers. *(3rd and 6th chakras)*

Blue - Blue opal is found in the center of thundereggs from Oregon and Idaho. Depending on the depth of blue, this opal impacts mental, psychic and speech areas, making it one of the easier opals to wear. It is more calming than other forms of opal. *(5th, 6th and 7th chakras)*

Cherry - Found in small rhyolite nodules, cherry opal is usually from Mexico. The color quality of cherry opal can vary in intensity, however it usually impacts the first, second and third chakras providing energy, determination, motivation and stimulation. It increases estrogen output, so individuals not needing estrogen may want to avoid this stone. It tends to increase blood flow which aids in reducing numbness and neuropathy. *(1st, 2nd and 3rd chakras)*

Hyalite - Hyalite opal occurs in small wads from Mexico. More elongated and clear stalagtite growths are found in Australia. The clarity of hyalite opal supports reducing confusion and anxiety, and the bubbly surface is calming and relaxing. It is a good mineral for people with Alzheimer's disease and for hyperactive children, because while it modulates nervous energies, it stimulates clearing of the mind. *(7th chakra)*

Girasol (Madagascar) - Girasol is a water-clear form of opal with pink iridescence which is found massive in Madagascar. It can be carved into various objects or made into spheres. Called the "Blessing Stone" metaphysically, girasol concretizes and makes real prayers, wishes and requests to spirit through the receival of blessing. It holds frequencies of angelic light and tone which respond well to even unspoken needs and longings of the heart. Its properties are so acutely refined that it is an excellent choice for Etheric Opener in the Spiritual Apothecary. *(7th chakra)*

White Fire (Australia) - "The Wish Stone." White fire opal is the stone of the Goddess of the North, bringer of dreams and wishes, spiritual fire and passion. This stone is intensely magical, and is the mineral precursor to white swan energies. This form of opal is prized for its ability to hold the purity of spirit during the dark night of the soul, and is an excellent companion for those in a period of deep spiritual inquiry, questing or crisis. White fire opal supports metamorphosis and transformation. *(7th chakra)*

OPALIZED WOOD: The qualities of petrified wood and opal are combined in this rock which is brittle and softer than agate. This material works with old knowledge banks (collective consciousness) and emphasizes awareness of the common earthly realm in order to view it clearly and without denial. This material tends to fix human traits in present awareness and emphasize it, whether it be mountain or valley behavior; therefore, it is more productively used in the living environment of those on an awakened path. It is not a good medallion or pocket stone for the non-aware. *(Lower chakras)*

ORPIMENT: Orpiment is arsenic sulfide, made up of two arsenic and three sulfur atoms (see Realgar). Do not touch this stone and then the body. Wash hands first. This is actually good advice for several types of stones which are made of poisonous elements such as arsenic. When orpiment is placed in the environment or focused on, it is a third chakra activator. It is highly energetic, stimulating and arousing. This is the masculine essence of Serpent Fire, and activates the right channel of kundalini flow up the spine. It is often found with realgar, and rapidly alters to realgar due to atmospheric exposure. See Proustite. *(3rd chakra)*

ORTHOCLASE: Orthoclase is a monoclinic potassium aluminum silicate classified in the feldspar group of minerals. It is dimorphous with microcline which forms in a triclinic crystal shape. Orthoclase is often white or colorless, but it can be colored blue, gray, pink or yellow. Some forms of orthoclase possess enough schiller luster (bronze iridescence) to be used as gemstones, but orthoclase is more useful industrially when mixed with kaolin and quartz to make porcelain, ceramic glazes, dental products and opalescent glass. Twin crystals of orthoclase are commonly found. The rounded nodule style, as well as many others, is a personal use stone, being specifically useful for centering and balancing into the Aum state of meditation. This is the ordered metempsychosis, the passing of the soul at death into another body, either human or animal. It can be used to enhance conscious death and shapeshifting practice. *(Auric chakras)*

OYSTER SHELL (Mother of Pearl): Like pearl, the softened, rainbow luster of mother-of-pearl works on the entire body, but principally the third eye area. Creativity and intuition are enhanced by wearing the shell pieces. *(6th chakra)*

PAUA SHELL: Paua is extremely physically energizing and stimulating. The ability to intuit physical information and signals is enhanced. It enhances the bond between physical and virtual senses. Virtual senses are psychic analogs to physical senses and are the points which resonate, allowing extrasensory phenomena to be translated to the physical. The integrity of paua shell is diminished by dyeing, allowing it to be worn safely by most individuals. Dye-free, paua shell should be worn only as medicine for the duration needed or by individuals who resonate to paua shell as a power totem. *(6th chakra when blue)*

PEACOCK ORE: See Chalcopyrite.

PEARL: Pearls grow from small grains of sand or other impurities layered with iridescent calcium carbonate inside mollusks either in salt or fresh water. Traditionally pearl was said to cure bleeding, jaundice, epilepsy, melancholy, mental illness, toxicity, and lung and eye problems. It was considered an aphrodisiac and a symbol of love and marriage. Pearl improves the digestion of events, often bridging between shock and acceptance. Therefore, it is applicable to post traumatic stress disorder. It reawakens lost innocence, and often encourages the wearer to focus on manifesting inspiration which has been dormant and even forgotten. It focuses attention and intention in a balanced emotional manner. It reduces irritability, and is especially soothing in combination with jade. Though it can be worn harmoniously anywhere on a person, it has a special link to the third eye area as it enhances visionary sensitivity as well as ecstatic vision. Pearl enables the wearer to observe the behavior of others and make self-loving choices about social contact.

Pearl is an excellent example of becoming what you focus on and choose to do now. It activates staying conscious of choices to change from ugly to beautiful, irritating to pleasing, rough to smooth. All these attributes allow pearl to increase the likelihood and manifestation of successful, loving relationships. Because pearls form around an irritating grain of sand, they represent the spiritualization of matter, spiritual maturity and the evolution of consciousness, and because they are hidden, they represent the search for "pearls of wisdom." Pearl is another gemstone said to result from lightning. The pearls of great price are guarded by aquatic dragons. Pearl is associated with the moon, and can be called "Cynthia's stone," the gift of the moon. It possesses feminine energy. Persia is the culture which has most highly regarded the pearl, and they say that pearls result from celestial semen, a drop of rain, falling to the ocean to become a pearl. They view pearls as symbols of virginity, and they decorated the crowns of kings with pearls. Pearl is the state gem of Kentucky and Tennessee. Pearl is the June birthstone. *(6th chakra)*

PECOS "DIAMOND": Pecos diamond has an elastic nature, being able to stretch out to accommodate the needs of the wearer as events unfold during a day. This quality allows the vibratory effect of pecos diamond to be compatible with a broad range of stones. It can be likened to a very diplomatic person. It brings out the best in the stones being worn at the same time. When used alone, it harmonizes the individual very quietly, very gently. It is a wonderful calmer for hysteria and anxiety, especially when induced by stress such as shock or trauma. In mental patients, it brings a quiescence which allows them to utilize their minds to understand what has caused pain in the heart. Pecos diamond operates on negative expressions of emotion, allowing the user to perceive himself in that state with detached compassion, rather than judgment and guilt. It does not stop the expression of negativity, however it aids in allowing the individual to gain control over negative thought patterns, and then to change them through conscious choice rather than repeat them. This stone has an interesting relationship to the interface of the conscious/unconscious, the mental/emotional, the trained/autonomic responses. It activates the heart chakra, and draws recognition with loving acceptance into this area. It is a friend of kunzite. *(4th chakra)*

PERICLINE: Pericline is a variety of albite often occurring as elongated, twinned crystals. Albite is a sodium aluminum silicate classified in the feldspar group. Pericline acts to dissolve or reduce by breaking down cellular structure, so it helps to remove tissue from scars. Toxin cleansers such as amber or malachite help to move particles from the body through the elimination system.

PERIDOT (Gem Forsterite): See description of Olivine. Forsterite is magnesium silicate forming in an orthorhombic crystal shape. In the gem industry, forsterite is called peridot. Peridot strengthens the physical body, and aligns and balances the subtle bodies. It aids in regeneration of many organs and glands: liver, pancreas, heart, spleen, adrenals and thyroid. It is superb as a mineral which alleviates depression and sorrow especially resulting from childhood trauma, abuse or experience of loss. Peridot is one of the few gemstones resulting directly from lava flows or igneous rocks, so it bears the ability to cauterize emotional wounds to allow the strength for deep healing to occur. It is excellent when combined with jade and pearl. Peridot is the August birthstone. *(4th chakra)*

PERISTERITE: Peristerite is a white to cream color variety of albite which is a sodium aluminum silicate listed in the feldspar group. It is a good beginner meditation stone, especially for children, as well as a sleep enhancer. For those lacking connection to the child within, this mineral provides support to be in touch with and visualize the internal child. It stimulates the imaginative center of the brain as well as the unconscious. It promotes lucid dreaming. *(6th chakra)*

PETALITE: This lithium aluminum silicate is a rare transformer, and as a clear gemstone is unsurpassed in dispelling depression, anxiety, hysteria and hallucinatory dissociation. It provides light instillation, meaning light is imparted gradually in metered amounts so the physical body is not overwhelmed or deprived during recovery. This mineral is unusually sensitive to human chemistry and energetics. It is an excellent base mineral for tinctures. *(7th chakra)*

PETERSITE: "King's Mineral." This mixite group mineral contains copper, potassium, phosphate, calcite, mercury, yttrium and thallium. It is colored mostly a rich, deep and flashy blue with golden/yellow mix. Petersite is highly susceptible to sound induction, and resonates with the second, fifth and eighth chakras. For people who operate from these chakras, petersite is a strong support mineral. It brings together the ability to make and pass judgments and to put them into action. It is excellent for people acting in a judgment capacity over the rights and destinies of others, as it resonates with compassion, strength, clarity and truth. An excellent companion stone to richterite (blue sugilite) and vanadinite, because it provides grounding and practical matter comprehension influences. *(2nd, 5th and 8th chakras)*

PETOSKY STONE: Petosky stones are fossil coral stones found on Lake Michigan's shore, and Petosky stone is the state stone of Michigan. This coral can help individuals to connect with the past and past lives. Petosky stone is for dreamers and visionaries. It has many places to store ideas and imaginative thoughts, and its structure provides companionship among great thoughts. Sometimes great ideas die for lack of enthusiastic support. Petosky stone aids in overcoming a sense of hopelessness and disillusionment with personal goals and dreams. *(6th chakra)*

PETRIFIED WOOD: Petrified wood is an agate replacement of the cellular structure of wood. Deformation at the microscopic level occurs when minerals invade, but the original wood grain pattern is often quite evident. Several geologic processes cause petrification, the two most common being lava flowing over trees and trees being inundated by mud. The long hardening process turns wood into a composite of jasper, agate and other minerals, typically in shades of brown, but red, orange, yellow, white and green are not uncommon. Petrified wood is the State gem of Washington and the State stone of Mississippi. Petrified wood is a vitality and disease protection stone. Its principal characteristic — great age and solidified change — helps one with past life recall during meditation. *(Lower chakras)*

PHACOLITE: Phacolite is a zeolite group mineral related to chabazite. It relieves stress, tension, nervous stomach, and feelings of anxiety. It promotes circulation and cell regeneration, and regulates body fluid balance. It is useful to counteract effects of chemotherapy. It is a good companion stone to orthoclase, talc and opalized woods. *(2nd chakra)*

PHENAKITE: Phenakite is a beryllium silicate. Phenakite activates crown chakra receivership and base chakra movement. It stimulates red and white healing conduits in the body stemming from the left thumb (white) and left index finger (red). It is an excellent physical integration mineral; good for choreographers, dancers, coaches and athletes. It helps raw, potential energy to be translated to kinetic, physical agility as well as creativity. Its ability to help individuals access etheric energies is unparalleled, however phenakite possesses a wild, untamed and unleashed quality which needs to be taken into account when wearing it. Imagine the mesmerization of the cobra by the serpent master, and you have an apt picture of the type of mastery of energy required to utilize phenakite to its highest potential. For people with pent up and repressive tendencies, phenakite provides stimulation to accept spontaneity, surprise and last minute fun. For people who are not known for their caution, phenakite may be too stimulating and may need to be avoided. Phenakite is excellent in a talk therapy capacity as it eases the ability to tell one's story, to let go, and to be willing to accept the unknown and uncertain aspects of life. It is applicable to people who have been repressed. Phenakite is an Etheric Opener in the Spiritual Apothecary. *(1st and 7th chakras)*

PIPESTONE (Catlinite): This prized red stone is still found in layers deep in a few select quarries in Minnesota, where Native Americans came from widespread areas to find it. Native Americans considered pipestone to be the sacred flesh and blood of their ancestors. When carved into pipes and other special ornaments, it was used to sanctify and solemnize all significant events. Its general use for pipes by many diverse tribes speaks highly of the Earth spirit connection it provides. *(1st and 4th chakras)*

PLATINUM: Platinum is a highly conductive metallic element. It is characteristically gray. Platinum moves energy to the base chakra to relieve constipation. It helps to let go of heavy depressive feelings which tend to dampen the spirit. When worn, this metal is particularly empowering for feminine energy. The high chakra (eighth) connection is mirrored from the first chakra. *(1st and 8th chakras)*

PLATTNERITE: Plattnerite is a lead oxide classified in the rutile group. It has an affinity to machines, and likes to be around machines such as cars or kitchen appliances, or even large ones such as electrical generators. The matrix of this mineral has natural properties which help the machine to understand its role in the natural and human worlds. Machines have terrible complexes about how they do not belong or fit in, and plattnerite is an exceptional ombudstone to have near so the "owner" can tune in and understand what may be wrong or needed by the machine. *(1st chakra)*

POP ROCKS (Boji Stones ™): The term Boji Stone™ has been trademarked and sold as a specially treated and more powerful version of this mineral. We will use the more generic term of pop rock. This iron pyrite concretion is a bridge between the mineral and plant kingdoms. It can provide subconscious training to humans wanting to bond with nature, specifically plants. Therefore, communication is the theme, especially telepathic.

PORPHYRY (Flowerstone) - Diorite, Granite and Quartz Varieties: The varieties range in color from gray/green to pink, red/violet, and purple. The darker red to purple was used in ancient Egyptian times and in the Renaissance by the English and Greeks as a talisman for leaders because it presents an enduring eloquence, and soothes the nervous system which can help with public duties, such as speaking and presiding over groups. All colors portray the interconnection of Earth's kingdoms. *(Chakra varies with color)*

PRASE (Green): This green mineral with white and brownish red jasper inclusions, from Australia, is a foundation-making mineral working well between the first, fourth and seventh chakras. In its greenest state, it resonates with the heart of the Great Mother. With inclusions, this mineral strengthens earth and sky connection for spiritual and/or meditative work. It is exceptionally well balanced. It is a phase-tone rock. This means that the rock is something like a cusp in astrology. It falls between two types of rock, primarily schist and jasper. It is an amalgam of both, but having the entire qualities of both, so it is unique. Green prase is an excellent "in-bath" or "in-water" stone to balance discordant frequencies. It is an excellent healer's stone for this reason. The healer can place the green prase at the solar plexus or heart chakra, and obtain information regarding what is imbalancing the patient. Green prase is excellent to place near newborns while they recover from the trauma of moving between extreme life phases, and this applies to dying individuals as well. The qualities of this mineral enable "presence of mind" to be experienced as a generalized or localized body knowledge so that the energy nuances of life (existence) can be read more accurately over a larger mass of tissue than the brain center. *(1st, 4th and 7th chakras)*

PREHNITE: Prehnite is a calcium aluminum silicate. It forms as a green to yellow botryoidal skin. Prehnite is considered an ascended mineral, in the context of off-planet work facilities. It has a strong ability to support life essence over great distance, so it is fine to hold during meditative travel and out-of-body work. It is not at all a grounder, so if grounding is an issue, remember to place obsidian (any type) at the knee or feet chakras when meditating. Lime-green prehnite is more of an uplifter. It is deceptively durable, and the quality that this evokes is to support the strength of the feminine principle, which is deceptively durable, and in the physical condition can long outlive its masculine counterpart. It is not a feminine mineral, per se, but its overall structure adds perspective to the yin force. Space travel is often perceived as a masculine occurrence, one reason it took women so long to get "out there"- in the traditional NASA form. However, the yin strength of space travel is that it requires no outstanding vehicle or propulsion. It is accomplished through the feminine power of knowing and shapeshifting. The ascended aspects of prehnite enable humans of both sexes to more fully capture when the knowing is present and real, so that it can become an embodied skill rather than a disembodied higher self quality. *(High range, upper chakras)*

PROUSTITE: Proustite is a silver arsenic sulfide. In its less evolved form, proustite is a first chakra mineral with properties of aiding blood and fluid flow in that area. In its high crystalline form, proustite is an unsurpassed mineral for healing, cleansing and balancing the first chakra. It is a powerful aid to kundalini practice as it prepares the three kundalini conduits with proper energy prior to release up the spine. In unison with crocoite, proustite supports the communion of tantric sex. Proustite represents the union of fire and serpent energies brought together in the light of the moon and the cleansing safety of silver. It is the feminine embodiment of the Serpent Fire. *(1st chakra)*

PSILOMELANE: Regarded as a colloidal manganese dioxide, the black/silver banding and swirls of psilomelane are characteristic. It looks much like hematite (iron oxide), but it is not. It is a hydrous barium magnesium oxide. Metaphysically it has similar grounding, energizing and protecting properties as hematite, but it can also be compared to wad and romanechite. Psilomelane is much more accessible than hematite and not as energetically accessible as wad. It is soothing and gentling, tending to greatest impact on the colon and intestines. *(1st and 2nd chakras)*

PURPURITE: Purpurite is a manganese iron phosphate. It is found in Colorado, South Dakota and Africa. The chalky yet iridescent quality of purpurite attests to its ability to act as a physical filter and mirror to determine if physical healing work is progressing on course with individual needs. Purpurite should be used as a "scope" to determine if leads to health improvement are actually being employed in the body. To use purpurite, place it above the crown chakra approximately three to four inches. Incoming light rays are filtered through the scanning network of the purpurite, and information is mirrored back to either the healing practitioner or the individual, depending on intuitive clarity and sensitivity. Purpurite is a useful tool for psychics as it allows balancing and re-energizing of the crown and third eye chakras after psychic work. Purpurite is also a good meditative tool as it allows full body frequency networking to occur. This means that the polarity of the individual is synchronized to allow the whole body network to function more serenely. It can be used to strengthen the impact of other healing stones being used in crystal layouts. For this purpose, purpurite is placed above the crown chakra, about two to three inches, while the individual lies down during the layout procedure. Purpurite possesses a frequency modality which is highly Atlantean, and allows much cross-connection and cross-reference work to be carried out associated with that era. This means that purpurite, in meditation, will allow the Atlantean experience to flow more deeply within the individual, so that knowledge and skills possessed during that existence can be utilized. Purpurite combines well with iolite. Purpurite is a Healer in the Spiritual Apothecary. *(Upper down to 6th chakra)*

PYRARGYRITE: Pyrargyrite is a lead antimony sulfide which can be compared chemically to proustite. Called "ruby silver," it is found associated with proustite. It is a first chakra mineral which complements proustite by aiding cellular level transfer of chemicals in the body. This transfer allows the nervous system to conduct kundalini energy with less stress and strain. It reduces energetic friction at the cellular level. It provides grounding which proustite lacks. It is physically protective. *(1st chakra)*

PYRITE: Pyrite is iron sulfide. The ancients used this mineral as a fire starter and as a mirror. It is called "fool's gold," because it is often mistaken in the field by novices as gold. Iron pyrite will give off sparks when struck with another stone. It stimulates the entire cardiovascular system, and its particular impact on the throat area hints at its assistance with communication and a friendly exchange of words. It bears a natural symbolic connection to the sun, especially in the "dollar" variety. It is a mundane embodiment of spirit energy on the physical level, and is often useful for children who are easily connected to the spirit realm, because it enables them to make stronger physical connection. It is normally the first stone preferred by metaphysical novices for grounding spirit energy to the body, however its simplicity makes pyrite perennially applicable and useful. *(1st, 4th, 5th and 7th chakras)*

PYROLUSITE: Pyrolusite is a manganese dioxide classified in the rutile group. Pyrolusite enhances connection to mechanical design and the underlying symmetry of interconnected networks of machines and technologies. It is an excellent stone for physicists, astronomers and astrologers who will "understand" it immediately. It enhances brain wave activities, making tasks done by these people flow easily, and strengthens comprehension of complex abstract ideas. *(7th chakra)*

PYROMORPHITE: Pyromorphite is a lead phosphate chloride classified in the apatite group. It bears a somewhat botryoidal surface with a grainy texture.

Golden - For any individual who suffered the abuse of their personal power and boundaries, this mineral reactivates the closed second chakra and soothes fears of invasion or loss of self. It stimulates self-worth and acceptance. *(2nd chakra)*

Green - Green pyromorphite is a quickening mineral, attractive to devas and sprites. It aids in communication with plant kingdom allies. Placed in a protective pagoda outside, it can be an important mineral among several in establishing rapport with local spirits. *(3rd and 4th chakras)*

Orange - Pyromorphite is also a quickening mineral, excellent for rainmakers and thunderbeings. It relates well to lightning. It is good for slow people with low blood pressure, and is not recommended for use by people with heart trouble. *(2nd chakra)*

PYROPHYLLITE (With Red Rutile): Pyrophyllite is a hydrous aluminum silicate. It is a clay mineral occurring in metamorphic rocks, resembling talc with radiating sections. This mineral stimulates the base and crown chakras. While held in the left hand with thumb and index fingers brought together during meditation, this mineral activates polarity balance. *(1st and 7th chakras)*

PYROXMANGITE: Pyroxmangite is a manganese silicate. It is a deep pink or fuschia colored mineral from Brazil. Pyroxmangite is a mineral of ancient and magical powers which brings in the energy of the ancestors, most particularly, grandmother Spider, the Great Weaver. It instills purity of vision and intent. It opens the heart chakra, and merges the physical heart with divine love. It offers vibrations of nurture, unconditional love and forgiveness. It will aid recovery from heart attack and all illnesses which stem from a diseased, weakened or broken heart.

It strengthens bone marrow and aids illnesses resulting from red blood cells, blood and bone marrow. It tends to resonate most strongly with the auric field, inducing changes there which translate to the body core over a period of weeks. It energizes the crown chakra, because the auric field showers waves of energy through the crown, heart and root chakras when using this mineral. *(1st, 4th and 7th chakras)*

PYRRHOTITE: Pyrrhotite is an iron sulfide which contains variable amounts of dissolved sulfur. Pyrrhotite expresses itself as a promoter of differentiation. It is a positive mineral to have around during the early weeks of pregnancy as it fosters healthy differentiation of cells. The magnetic qualities of this mineral also relate to emotional issues of differentiation such as the development of separation and leaving home. It helps to broaden the awareness and acceptance of differences in others. It is a natural embodiment of masculine/father spirit/sky and feminine/mother spirit/earth energies. It can be thought of as earth lightning. *(2nd, 3rd and 4th chakras)*

QUAHOG (Wampum): Quahog is a white and purple shell of wonderful character which stimulates creativity and strengthens both clarity and intent. It is called the "D" shell, because it is an unparalleled aid to the creative person needing assistance with drive, direction or discipline. *(6th chakra)*

QUARTZ: Quartz is simply silicon dioxide, the most common mineral on earth. The word quartz derives from *kwardy,* in Slavic, meaning hard. Quartz rock has been found with the remains of prehistoric humans dating back 75,000 years. The word crystal derives from the Greek crystallos which means ice. The clear portion of quartz is a universal tool described in many books for an abundance of technological and metaphysical purposes. Only quartz and tourmaline lack central symmetry in their crystal structures which makes them piezoelectric. Pressure applied across opposing prism faces allows the build up of an opposite charge, and releasing pressure reverses the effect. This shared property has made quartz, in particular, and tourmaline important to electronic applications. Quartz is the state mineral of Arkansas and the state gem of Georgia. All forms of quartz are listed in the Spiritual Apothecary in the Healers division. Though many of the minerals described in this book have a quartz or silicate base, the types of quartz listed here are those whose common names are associated with quartz:

Amethyst - The purple color is caused by microferric iron inclusions in clear quartz. See Amethyst listings.

Ametrine - The gods must be lucid! This combination of amethyst and citrine is one of the great marriages of the mineral kingdom. Not only are the two colors of orange and purple revered in India for their sacred qualities, but these colors represent the middle finger of each hand. Orange applies to the left hand and purple to the right. The meridians of these fingers are activated by ametrine enabling left and right brain coordination and deep meditative connection between the second and sixth chakras. Ametrine builds space like its going out of style. It enlarges and makes greater everything around it by drawing on the natural abilities of the two chakras it activates. This is a mineral of sublime balance and cohesion, bringing together in form what can only be called perfection to the senses. It is the marriage of lower and upper, creativity and wisdom, sky/heaven and earth/ground, being and doing. It embodies the ability to manifest healing and health through the union of mind and body (matter), the body/mind connection. Ametrine is a feminine power mineral as it links two feminine chakras. *(2nd and 6th chakras)*

Blue (Gold Gas) - Also called "Aqua Aura®." An immensely powerful protection combination which fuses the white, blue, and gold rays into an iridescent display of light. Excellent combination for meditation and adding strength to healing work. An all around useful tool, because it will enhance all interactions. *(All and auric chakras)*

Blue Quartz - The natural blue/gray color of blue quartz is caused by inclusions. Blue quartz is excellent for optical stimulation of distressed or traumatized tissues. It is an excellent exerciser for those wishing to eliminate glasses, and provides good overall toning when used routinely. It is suggested that the specimen be placed in a lighted area when it is back lighted so that the individual obtains benefits from light refracting through the piece. *(5th to 6th chakras)*

Blue "Rutile-like"-included Quartz (actually Tourmaline) - There has been much confusion over these inclusions, called crocidolite, riebeckite, rutile and "blue lightning," but the Smithsonian tested it as tourmaline. The "Law Giver" stone; extremely refined energies are transmitted through this special rutile crystal. It is called the "Law Giver," because it transmits accurately and precisely universal principles and laws. A crystal with a triangle on a terminated face is called a "Law Keeper," meaning that the interstitial spaces are capable of storing universal codes and information. *(6th and 7th chakras)*

Castle-tip - Castle-tip quartz acts as a lightning tool or laser diffusion rock. Castle-tip quartz is a supreme healing tool, because the focal impulse is released to the body through the many tips allowing the strength of lightning to be translated to a wider expression something like water from a shower head. One thin stream of water cannot cleanse in the same manner as a wide spray. Excellent for removing layers of fear and anger and the attached negativities of others. A forgiving expression of quartz in that the beam diffusion is gentle and generally easily felt by practitioners, so work flows effortlessly. The distinctive multi-terminated end is caused by contact with titanium dioxide. *(7th chakra and universal)*

Cathedral (also called Water, Skeletal, Jacare or Elestial) - The smoky quartz ridges on the faces pull in Earth energy and combine it with the clear quartz etheric energies to provide a strong and balanced effect. Since water is often found trapped in these crystals, it enables magical, transformative flow as well. The many positive aspects of these crystals may be summed up by the keynote traits of connectivity and balance. *(1st and 7th chakras)*

Celestial - Celestial quartz is a milky-white phantom quartz from Arkansas with deep etchings in alternate termination faces. These etchings allow for greater energy exchange. Celestial quartz is worn near the collar bone as the amice is worn in ecclesiastical vestments. This placement strengthens the shoulder carrying capacity of white light and pure vibration and tone. The proper crystal for an individual is tuned to the body over a period of a year and specific purposes of the individual's call to service are made clear during this time. *(7th chakra)*

Chlorite-included - Chlorite is a green mica inclusion in quartz, not uncommonly included with other minerals which cause other colors. Programmed or not, this form of quartz is an exceptional healing tool related to the entire ingestion, digestion and elimination cycle. When activated by rubbing vigorously between the palms, it can provide healing to ulcerative problems. One reason this crystal is nicknamed "the potato crusher" is that it breaks down carbohydrates, and allows proper assimilation in the body. It is an excellent crystal for vegetarians or plant lovers in general, and when programmed and placed in the garden, they are a constant reminder to the plants of your love and appreciation for them.

Chlorite-included quartz is an Etheric Conditioner in the Spiritual Apothecary. It is singled out for notice as a specific member of the apothecary, because it is uniquely endowed with the ability to send healing frequencies without programming. It is a natural healing tool which is enhanced by human interaction, but it isn't required. Passive presence in a room or jewelry affords substantial benefits. It holds a position of authority in the quartz realm as a specimen which can be consulted specifically for healing and diagnostic information. Healing practitioners may find that it is an essential component of their healing repertory. As a psychic tool, chlorite-included quartz is uniquely endowed to transmit the Green Ray which is simultaneously a physical and etheric realm interface. It is said internal angles within the Great Pyramid are called green rays, and these angles are responsible for maintaining the viability of seeds for many millenia. The Great Pyramid is called by Stan Tenen a conduction device for shapeshifting and time travel. Chlorite-included quartz is an important element in achieving physical expression of both these concepts. *(4th and 7th chakras)*

Citrine - Citrine is quartz included with colloidal iron, which gives citrine its orange to yellow color. It can be artificially manufactured by heat treating natural amethyst. In its yellow expression, citrine increases self-awareness and psychic awareness originating from body receptors. As it works to support the third chakra, control of mental and emotional stimulus is increased, because more energy can be focused into these areas. Because the third chakra is where we inhale our inspiration for life, citrine facilitates influx of white light and positive vibrations through the solar plexus. It aids in identifying and manifesting life purpose. In its more orange form, citrine enhances digestion, and improves circulation of blood throughout the trunk and legs. It beneficially affects all the organs of the second and third chakras region, including heart, digestive, liver, kidneys, spleen, appendix, colon and gall bladder. It is a powerful creativity support as well as will force strengthener. It can improve financial affairs by stimulating astrological interests and abilities of the natural zodiac's second house, and supports expression of life force energy.

Citrine is a remarkable quartz family member, for two reasons. It is not especially rare, but mineralogically, yellow is not easily obtained, so its ability to assist with second and third chakra issues makes it valuable. Secondly, it is found with amethyst, at a site in Bolivia, called ametrine, creating a cross-planet correlation to the Indian reverence for purple and orange as sacred colors. Citrine is an Etheric Sealer in the Spiritual Apothecary. *(2nd and 3rd chakras depending on specimen)*

Clear - Clear quartz is unadulterated silicon dioxide, and is the most common variety of quartz as well as one of the most common minerals in the Earth's crust (twelve percent by volume). We characterize it as man's best friend in the mineral spectrum, therefore it embodies the energy of dog. The link of oneness it provides throughout the earth gives it its greatest significance as being a stabilizer of divine forces on the physical plane. It brings spiritual consciousness to the material world. It can be used for all manner of healing work and therapies, and quartz acts as a medium for meditation, clairvoyance, wisdom, healing and communication with all levels and kingdoms of life. In many indigenous cultures, the light body of quartz is integral to the initiation and work of shamans. Quartz is used universally as a tool of divination and magic.

Its balanced polarity and properties in common with many of our body cells and substances make it an unparalleled aid to our physical, emotional, mental and spiritual wellbeing. This is the best mineral for affirmation-programming work as it is the most universal form of crystalline structure. It aids connection to the higher and inner selves as well as the Source (Light), spirit guides and angelic mentors. Quartz receives, stores and releases energy which is why it is so useful in modern communication technologies. It is both piezoelectric and pyroelectric (except in Brazilian twins where the charges of the twins cancel each out because they are oriented in exactly opposite directions), and is used to create solid state fields of energy. Clear quartz has revolutionized human affairs and civilization. It is now so important it is lab-grown, but healing practitioners generally regard natural quartz as a more sophisticated tool.

Clear quartz resonates to qualities of loyalty, fealty, equality and "one for all, all for one." Understanding the properties of clear quartz brings new light to the Biblical injunction, the meek shall inherit the earth. Those who receive and let flow through them the Divine Light in complete and utter humility will flourish on earth, and will literally glow with goodness and light. Clear quartz, the simplest of crystal and mineral forms, urges a deep wonder within and subsumes itself in humility to wide open acceptance of the Source. Light (the Son) striking crystal (Mary) has long been the symbol of the Immaculate Conception.

Quartz is the lowliest common denominator chosen by universal intelligence to be the prime medium of light and information on earth. Its lesson to us is its divine simplicity, its elegant grace and humility in service. In its mundane abundance on Earth, quartz becomes meek, the true inheritor of Earth. In this age of modern technology, we have begun to see widespread destruction of earth's habitats, pollution and over-consumption of resources. Some of us know that humans are expendable, and feel the hot breath of extinction blowing on our necks just as we annihilate so much around us. The truly detached observer is silicon, waiting for the era of carbon-based life form dominance to wane. *(7th chakra)*

Dragon's Tooth Diamontina - A terminology for "laser wands" with a swirled-tooth root look. Good for dowsing. Excellent as a diagnostic tool for body applications. Especially useful for energy applications. *(7th chakra and universal)*

Drusy - Drusy is a mineralogical term applied to small quartz points of even height under 2 mm. They emit a light and lively energy. The cheerful energy is attractive to children and the inner child. Placing jewelry or other stones on drusy is good for energy cleaning and balancing. Drusy is affected by the mineral it covers, if not quartz. *(7th chakra and universal)*

Fisher Mountain Blue Phantom (Blue/Black) - This form of quartz is characterized by gray phantoms of manganese. It is compatible with the herb comfrey. Good combination mineral to enhance poultice treatments. Both a physical and emotional "drawer" stone. Excellent for absorption work. Needs to be kept cleansed as casual handling of the stone stimulates its drawing capacity. Likes to be near the skin. *(All chakras)*

Geodes (Quartz) - Geodes are vugs in host rock which are lined or bound by a crust of agate in which forms clusters of quartz points, often clear, amethyst or citrine. Therefore, geodes are the parent or home of most quartz and any other minerals found within the "wholeness" of their domain. They symbolize the beauty and power found "within" us, and the completeness of the circle or sphere. The lore about how the abundance of Brazilian minerals was discovered includes a story about a German geologist walking in a remote region where local farmers used geodes as fencing material. He recognized what they were, and opened up a few. Upon asking where they came from, he was directed to regions which today are mined extensively for their mineral riches. *(7th and other chakras)*

Hematite-included - Hematite inclusions in quartz cause a black to silver and often reddish color. Hematite-included quartz crystals are a master quality crystal. The interaction of hematite and quartz produces a high quality visual transmitter and receiver, depending on the termination shape of the crystal. These can be used by clairvoyants (visual psychics) to counteract fatigue during readings, and to optimize accuracy. Held up to the third eye, at first darkness will occur, and then receivership begins, much like a movie. The same occurs during transmission. These crystals will be more easily and frequently used as more people become aware of and master higher self awareness and transcendental dialogue. As a healing tool, this crystal combination eases catatonia and autism, allowing the intervenor or practitioner to access the higher functioning aspect of the individual so self-healing can begin. *(7th through 12th chakras)*

Herkimer "Diamond" - See Herkimer "Diamond" Quartz.

Iron Sulfide-included ("Churgur") - This inclusion causes a pink to red colored garden, found often in Brazilian quartz. It is often polished, because the inclusions grunge up the surface. It is excellent for internal warming or cooling. Place in left hand of person experiencing shock to revitalize internal processes and bring balance to body systems. Fever is reduced by placing in right hand. Can transmit warmth to frostbitten extremities from the inside out to support the body's abilities to recover. A practitioner can do this for frostbite: hold the crystal in the palm with the little finger and place the thumb and other fingers near, but not on, the injured parts. Headache due to sunstroke is alleviated, and heat prostration can be avoided altogether by wearing this crystal. This is a healing tool specifically, and is not meant to be worn as an ornament as it has a tremendous impact on the internal homeostatic system. *(All chakras)*

Lavender - Similar affect as light amethyst but usually found massive like rose quartz. *(6th chakra)*

Midnight Black - Occurs as drusy stalagtites or small nodules. Properties are similar to schorl or black tourmaline, but with a spatial stimulation aspect. Activates the skull. Good for artists especially visual artists and excellent as a meditation tool. *(1st chakra)*

Orchid - Orchid quartz is a color of rose quartz which comes from top grade Brazilian or Madagascar sources. It is a gemmy pink/purple which tunes the heart for spiritual connection. *(4th chakra)*

Papagoite-Included - This is a form of quartz with a dark, powder-blue inclusion found in South Africa. Papagoite is a "Sky Door" stone, providing in high, protected form a mandalic energy which represents heaven/spirit and earth/physical complementary principles. Papagoite strengthens the personal center of power and the auric protective field. Papagoite is very similar to ajoite in function, and is an alternative to ajoite for those whose individual frequency does not resonate efficiently with ajoite or ocean picture stone. *(6th and 7th chakras)*

Phantom - Phantom quartz is caused by visible mineral tracings deposited in growth faces. They are often highly regarded by mineral collectors and metaphysical users alike. Quartz phantoms embody the principle of a seed idea within a fully formed concept. They also represent the ancestral basis of knowledge and the chain of physical learning and teaching which leads to the creative expression of knowledge and enlightenment. Quartz phantoms remind us that within the obvious, physical manifestation of life lies the mysterious, ineffable spirit essence describing that life. *(7th chakra)*

Pleiadean - Pleiadean quartz is characterized by busy, included faces of quartz like celestials, but tending toward greater clarity and originating in a different location in Arkansas. Pleiadean quartz is the "Language of Light" stone. This formation of quartz is readily detected as a life transformer as the energy moving through this quartz will impact an individual very deeply. Changes stimulated by working with this mineral may cause abrupt and cataclysmic upheavals. Healing crisis after healing crisis may occur for months, especially with larger specimens. The end result will be an obvious manifestation of working toward light goals despite not being clear what one should be doing with one's life. As these times turn toward embracing this alternative as a means for continuing life on Earth, one will find oneself already immersed in the path and vocation of calling. In a time when "clarity, integrity" is all you hear until you want to scream from the burden and responsibility of it all, hold this quartz to your heart and know that chaos is but an absence of form. While you hold the form of Pleiades intention in your hand, resonate with the purity which calls you home and brings you safety and comfort. This quartz is a new find in Arkansas, and looks much like celestial quartz discussed above. *(7th chakra)*

Prasio (with Hedenbergite) - A mineral prized by Ashtoreth (Astarte), the Syrian and Phoenician goddess of sexual love and fertility, as a healer of the first and second chakras, especially the reproductive organs. When combined with crocoite, prasio quartz enables tissue restoration to occur more rapidly. The two minerals in tandem produce a variety of results depending on the physical problem. *(1st and 2nd chakras)*

Red Phantom (Red/Brown) - This type is excellent for bronchial problems, tissue regeneration, and aeration of blood. It is good for wounds which bleed and ooze, helping clotting and reducing scarring. It is a tissue/blood interface stone.

Red Quartz "Balls" - These are the Spanish quartz multi-terminated pea to marble size spheres of quartz apparently with high iron content. *(1st chakra)*

Rose - The color of rose quartz is produced by impurities of titanium or inclusions of manganese. Rose quartz is the quartz of the heart, an emotional release and balancing stone. Its pink color resonates warmth and love. It helps all types of lonely and deprived souls. It is especially good for children and those who needed love when children. The rose energy flows in with the circulation of the heart vessels. Rose quartz is the state mineral of South Dakota. *(4th chakra)*

Rutilated - Rutile in quartz is titanium dioxide needles surrounded by quartz. They are found primarily in Brazil. This form of quartz adds energy and life force to clear quartz properties for all ages. It acts as a regenerator of people, all animals, crystals and plants. It helps resist aging and the general cellular breakdown caused by time, and provides energy to the aged to counteract senility and dementia. Rutile is a significant celestial, off-planet connector which draws universal life force energy into the body. See Rutile. *(All chakras)*

Sceptor - A thin shaft and bulb-like head gives this variety more storage ability for energy before it is released (piezioelectrically) when touched. *(7th chakra)*

Siberian Blue - Siberian blue quartz is an excellent achievement in the lab-grown category of minerals. Cobalt is added to create the rich, blue color. The deeper cobalt blue variety is an aura and sinus cleanser, and helps to counteract heat illnesses and anxiety. Because it is manmade, its properties tend to simplify and become more applicable to many ailments, just as single alkaloids taken from plants become useful for a variety of problems. For instance, antihystamines are useful for anxiety control as well as reducing sinus inflammation. The icy, periwinkle color variety is especially tuned to fairy minerals. When combined with fairy minerals such as dioptase, releases of emotional stress are facilitated. *(5th and 6th chakras)*

Silver Rainbow (Silver Gas) - A silver coating is added to clear quartz in a vacuum, which leaves an iridescent pink color without completely blocking the transparency of the quartz. This is an excellent mineral combination for supporting the realization and manifestation of dreams and aspirations. It is useful for any person having to audition or compete for recognition, and melds easily with all body frequencies. *(6th and 7th chakras)*

Smoky - Smoky quartz is the result of ultraviolet radiation turning aluminum to a dark brown or even black. This can occur naturally or humans can treat the specimens to produce the same result, however when manmade, the appearance is often distinguishable. The quartz looks as if it has been fried, and these treated samples require a great deal of loving energy to rehabilitate. Natural smoky quartz assists the entire lower body in clearing and cleansing physically and emotionally. Its appearance is symbolic of its ability to bring white light down through the body to the base of the spine. It provides grounding energy for successful action, and naturally stimulates the adrenals. Smoky quartz is the state gem of New Hampshire. *(1st chakra)*

Solution - A "fairy garden" mineral. Excellent for gem elixirs. A companion crystal to dioptase and anhydrite. When creating a chakra association layout for young children, solution quartz relates to the crown chakra. Blue/purple sugilite to the third eye. Anhydrite to the throat. Dioptase to the heart. Apatite to the solar plexus. Carnelian or Botswana agate to the womb chakra, and corundum crystal to the base chakra. These are the fairy minerals that creatures and children know and understand. *(7th chakra)*

Star - This is quartz with asterism, or a "star" in it, caused by micro-inclusions lined up along internal growth lines. Four or six ray stars are most common. Energy is naturally focused at the star point(s).

Strawberry Rutile - Dense red needles make strawberry rutile, which is rare, highly desirable when polished for jewelry. Strawberry quartz is a mineral of immense elemental power. It provides a strong, energized base of material communication and association, allowing the transference of archetypal memory to the physical reality or plane. This mineral facilitates the consciousness of self to incorporate. The rutile in strawberry quartz reminds the psyche of past life work and lessons, while the red/amethyst interface of strawberry quartz empowers this transaction, stepping up the central nervous system to accept kundalini energy. The pulsations facilitate change and awareness of freedom. The mind conceives more readily unleashing new ideas and thought scales. While strawberry quartz is the great instigator, it does not generally cause immediately recognizable results. The completion of working with this mineral is trust and patience. The processes encouraged by this mineral must grow in their own synchrony and proper dynamic. Willingness and gratefulness of heart will supply the best soil for accomplishment of satisfactory results. This mineral impacts all body chakras. *(1st through 7th chakras)*

Tabular - Tabular quartz crystals are characterized by a flat or squashed type of growth which looks like a book. Called "Inscription stone" metaphysically, the name, tabular quartz, is apt. Tablets of quartz are what Moses brought down from the mountain. They were carved and etched just as described in the Old Testament, but they were surpassingly powerful because they had been programmed by Yahweh to emanate the Word into physical reality every moment of every day. Tabular quartz crystals are the books of the universe. Immense quantities of information can be transmitted across great distances in small packages. The science of encoding and retrieving this information awaits common adaptation, and is predicted by crystallographers to be achieved by 2005. *(7th chakra)*

Tangerine - Tangerine quartz has an orange coating which makes it rare. It is found in Brazil or Madagascar. Tangerine quartz is a feminine mineral embodying the dual purposes of empathy and right will. Its vibratory requirements of the wearer or user place it in a class equal with pink danburite and blue tiger eye (hawk's eye). It possesses a seasonal predilection for summer, and is best used in bright sun conditions. *(2nd and 7th chakras)*

Tibetan (Orange Healer) - Tibetans rub cinnabar on quartz carvings which nearly achieves the same effect as this natural quartz. This combination draws together earth and galactic vibrations as well as the second and tenth chakras. It enhances the opening of the second chakra for healing and creative work. *(2nd and 10th chakras)*

Yellow Iron-Included (Golden Healer) - This form of quartz coated and included with iron oxide is found in Arkansas. Called an "energy pack" metaphysically, because long distance runners, travelers and explorers can use this mineral to replenish depleted energies. Dispelling of accumulated, unwanted energies should first be done with fluorite or programmed, for this purpose, quartz. Many healing functions naturally occurring in the mind-body-spirit are boosted and enlarged in scope by "golden healers." This is an excellent mineral for toning of the solar plexus. This toning process deepens the absorption of life force and purpose energies, adding clearer intent to the inhalation of self from the higher self and spirit. It is quite conducive to harmonizing the third and sixth chakras through crystal gazing in candlelight. *(3rd chakra and etheric chakras)*

QUARTZITE (Twelve-sided Crystal): "C-Diamond." Mineralogically unnamed as far as we know. Found in the Dominican Republic. Apostolic energies flow through this crystal. Immense divine conductivity in a compact, perfectly shaped form. Colored opaque tan to white. *(7th chakra)*

REALGAR (Crystalline): Realgar is arsenic sulfide, made up of one atom each of arsenic and sulfur. It is deeply, pervasively energizing. For those slow to wake up in the morning, a few minutes with realgar will reduce caffeine dependence and increase mental clarity. Realgar strengthens chakra connections all the way up the spine, so it is a good tool for body work or bioenergetics. Realgar tends to load chakra and meridian circuits to the maximum, because it stimulates the central channel of kundalini. It is the balance of masculine and feminine energy, and is the prime mineral of Serpent Fire. It is not recommended for long term wear. Wash hands after use, and keep away from the mouth and children. *(1st chakra)*

REALGAR/ORPIMENT: These minerals, not unusually found together, activate vibrancy by stimulating the reproductive and digestive glands, promoting grounded action and kundalini flow. The reproductive system provides the necessary energy for constant kundalini emissions up the spine to the brain. It is interesting to note that affinity to the energies that realgar and orpiment afford can be nurtured and systematically developed. In the past, arsenic was taken slowly and tolerated by those attempting to poison others. Kundalini practice, to be survived, can be tolerated by the body as it acclimates to a deliberate process of metered flow. Use these minerals energetically. Do not ingest any portion of them. Wash hands after use. *(1st to 3rd chakras)*

RHODIZITE: Rhodizite contains potassium, cesium, aluminum, beryllium, boron and oxygen. This mineral expresses itself in pea-sized, cream-colored, opaque, garnet-shaped dodecahedrons. Rhodizite is a purgative for the bowels and a purifier of both the elimination and inhalation systems. When worn continuously, these systems will function more smoothly. Life purpose is clarified, but additionally, direction, discipline and stamina are all supported to achieve desired dreams, goals and aspirations. This mineral is an excellent combination of elements which is vitally useful to practitioners in many subtle energy disciplines from shamanism to research. It supplies the needed strength to withstand buffeting from constantly shifting energy fields as well as sterilization barriers to ward off unwanted microbes, energies and thought forms. *(1st through 4th chakras and subtle body)*

RHODOCHROSITE: Rhodochrosite is a manganese carbonate classified in the calcite group. It forms in light pink to fuschia and white bands, and is mostly found in Argentina (which is the Spanish word for silver). Rhodochrosite transmits high heart and emotional energies, and moves energies toroidally. This means that heart chakra energies move in a donut pattern up, to the side, down and back into the body in an unending flow. This accelerates and integrates higher energies, which vibrate at lower frequencies, with lower energies, which vibrate at higher frequencies. Rhodochrosite melds the upper triad (5th, 6th and 7th) of chakras through the heart (4th) with the lower chakra triad (1st, 2nd and 3rd). This melding mixes spiritual and physical energies into a new form, neither one nor the other, and the sum is definitely, clearly superior in experience to either experienced alone. Rhodochrosite is excelled only by manganocalcite in the ability to transmit loving tenderness. Rhodochrosite serves the subtle body nearly exclusively, while manganocalcite bridges to the bones. Rhodochrosite cleanses and amplifies the aura and other subtle bodies, and is listed in the Spiritual Apothecary under Auric Cleansers and Conditioners. *(2nd through 4th chakras and subtle body)*

RHODONITE: Rhodonite is a manganese iron magnesium calcium silicate. It works with the blood to transmit energies between the first and fourth chakras. It is a soothing mineral to the nervous system and balances the endocrine system. It supplies a detached frequency of feminine iron will, the proverbial iron fist in the velvet glove. It is a mineral which teaches the art of defensive energy, so it applies well to aikido, tae kwan do and other martial arts in the defensive tradition. It fosters heart strategy where all are encouraged to win and where competition is with oneself. Rhodonite is a mineral of discipline which feels appropriate, attractive rather than burdensome. It bears the stamp of Eastern consciousness rising from the earth and rocks of Asia and the Pacific Northwest of the United States. It blends well with jade in healing modalities as well as jewelry. Rhodonite is the state gem of Massachusetts. *(1st chakra)*

Massive - Black and pink to deep pink rhodonite reduces emotional imbalance as it blends the heart pink with the dark root chakra colors. It helps a person self-actualize as it increases awareness of dignity, self-respect and self-regard. Rhodonite grounds all these essential awarenesses to a stable foundation. Pieces with very little black in them improve the sense of hearing as this form of rhodonite is a sound-sensitivity mineral. This form is a singer's stone, and aids individuals to express their heart's desire, longing, quest or passion. It is a unique aid to insight as it strengthens the heart connection to the higher self, bridging the intelligence and learning ability of the heart tissue with the higher self while bypassing the brain. *(1st chakra)*

Crystalline - An excellent mineral form for healing male genital cancers, sexually transmitted diseases, and physical manifestation of rape or incest trauma. Crystalline rhodonite, with the following minerals for each chakra, is beneficial for full-body healing effect: crocoite, second chakra; lime green heliodor, third chakra; green apatite, fourth chakra; sugilite (blue), fifth chakra; quahog, paua shell, or amethyst, sixth chakra; and selenite, seventh chakra. *(1st and 2nd chakras)*

RHYOLITE: Rhyolite is a form of igneous rock with a felsic (alkaline feldspar) chemistry mixed with quartz. Biotite, magnetite, albite and ilmenite can be found with it. The background of the rock is usually very light with multicolored inclusion flakes, but it can tend to the totally black end of the spectrum which is known as obsidian. Pumice is rhyolitic volcanic glass. Rhyolite allows for the blending of communication skills, speaking, listening and being heard. It ups the amperage on heart level emotional expressions, bringing in depths of self-awareness which enhances the ability to communicate clearly. Male and female polarities are balanced by this mixture of chemical elements. *(4th and 3rd chakras)*

RICHTERITE: Richterite is a sodium calcium magnesium silicate and is classified in the amphibole group. It is suggested for recovery from losses incurred by enemies or rivals. It deepens the ability to contemplate and meditate on the lessons of experience and to gain perspective which fosters acceptance and detachment from judgment. It increases the likelihood that an individual will arrive at a place of true understanding and forgiveness as well as love. It enlarges the capacity to discern, distinguish and discriminate what is true, honest, integral and righteous. This is the single most efficient mineral for reducing acid conditions in the body, thereby restoring alkaline levels, as well as addressing arthritic and rheumatic conditions. It is excellent for children to help with bones and teeth, and it is an age-defying mineral for the old, providing vitality and vigor.

This mineral is an excellent amalgam of elements to address nervous tension, stress, digestive disorders, and vitamin absorption. It can be considered the kelp of minerals. It is soothing to cells and tissues, and when complemented by amblygonite provides a well-rounded effect for hysteria, extreme anxiety, and intense grief. When combined with lazulite, symptoms of schizophrenia, bipolarity and borderline issues can be alleviated. When combined with either variscite or vivianite, symptoms of anxiety, compulsion, obsession, chronic fatigue, post traumatic stress disorder, and paranoia can be neutralized. When combined with turquoise, symptoms of catatonia, autism, dissociation, epilepsy and degenerative muscular or nervous system disorders can be relieved. Like other amphibole group minerals, richterite aligns with frog energy. *(6th chakra)*

RIEBECKITE: Riebeckite is a sodium double iron silicate classified in the amphibole group. It is a mineral of firmness, resolution, uncompromising values and integrity. It helps with boundary issues of all kinds as well as placing boundaries around appropriate times to express grief, tears and pain following loss. Its power resides firmly in its durability and dark potency. It stands unafraid, unshielded in the face of the unknown, and holds most profoundly the ability to go within the darkness to reveal new life. The cycle of the bear is quite instructive here, as the bear is a formidable animal who must retreat yearly into the cave, to wait and slowly starve in order to birth her young. She is unafraid and she emerges each spring with the reserves to carry on again. In association with eudialyte, this combination grounds, centers and uplifts. It is a powerful specimen for creating a safe and impregnable space for creativity to occur and for establishing new foundations and goals after loss. Riebeckite resonates to frog energy. See Arfvedsonite. *(1st and 2nd chakras)*

ROMANECHITE: Romanechite is a hydrated barium manganese oxide found coating some specimens of psilomelane. The dark, velvet texture of this mineral expresses a commanding gentleness, and is helpful in overcoming fear of the dark. It is useful for bringing people back to a safe place if work in the astral realm has caused problems. It embodies the essence of deer. *(1st chakra)*

ROSASITE: Rosasite is a hydrous copper zinc carbonate similar in appearance to aurichalcite. It is colored bluish to green. Rosasite is biophyllic, meaning that it loves living forms. It is the type of mineral which yearns for incorporation in life forms with greater degrees of freedom available to them. The idea that one can sink one's teeth into something in order to understand it or enjoy the challenge of working with it is exactly how rosasite affects the energy field. It induces enthusiasm, willingness and readiness to play, work or receive spiritual grace. It is excellent for the immune system, oxygenation of the blood, and overall well-being. Rosasite strengthens eye muscles, often improving eyesight. *(4th to 6th chakra)*

ROYALITE: Royalite varies from light lavender to dark purple, but with calcite and pyrite inclusions, it looks like the texture of lapis lazuli. It does not fluoresce as lapis lazuli does. It is a fifth dimension-oriented mineral, excellent for meditation, contemplation or spiritual attunement. It tends to "darken the playing field," meaning that it shields the crown chakra from distractions so that psychic images can appear as light on a dark background. It is a mineral which embodies the concept of "stream-of-consciousness," promoting awareness of one's interior monologue in order to make changes and choices in behavior. It is a mineral which aids spiritual awakening, times of spiritual crisis, and incorporation of Divine light into the body. It is not suggested for use in times of logical work or activities involving machinery. *(7th chakra)*

RUBY (Corundum): Ruby is the common name for red corundum which is an aluminum oxide classified in the hematite group. In cabochon shapes, some rubies exhibit a form of asterism which produces starring. Other colors of corundum are called sapphire. India is a prime mining location for rubies, and this area of the world has benefited from the effect of a fairly abundant supply of this gemstone. Next to diamond, corundum is the hardest mineral. Ruby's simple formula places it in a category of single energetic effect, and the aluminum-oxygen combination provides strength, resilience, and endurance. Ruby is a mineral which supports the heart as the focal point of heart-centered meditation leading to a variety of spiritual goals and fruits. Ruby strengthens the heart both physically and emotionally, and it focuses the ability to love. It provides large amounts of energy for vitality and vigor, and increases courage and physical stamina. It is a stone of masculine virility and fertility. It promotes emotional and physical stability, as it is a preserver of emotional and physical strength. It is a stone of raw, essential power. This is not a mineral which enhances feminine, receptive or lunar qualities.

Ruby resonates to Pegasus, or sky-horse medicine. It supports the elimination and digestive systems as well as the reproductive system. Ascribed to the month of July as a traditional gemstone, ruby applies directly to the solar power in physical form (Leo) rather than the lunar power in physical form (Cancer). The astrological qualities of Leo that ruby expresses include detached governance, inspirational and zealous leadership, dignity, fealty, loyalty, ardent love, passion and commitment. Ruby represents fire energy in the corundum family. In the past, it was highly regarded as a gemstone of royalty, authority, longevity and beauty. It is known to counteract Alzheimer's disease, and to bring energy to the elderly. When a ruby was placed in the third eye area of a dragon, it was known as carbuncle. Carbuncle was associated with determination, success and energy, as well as war and bloodshed. The second horse of the apocalypse is red, and his rider signifies the coming of war. Ruby represents many of the attributes of Mars, the red planet and the god of war. Mars is described as aggressive, energetic, desireous and fertile. Ruby was believed to resist and ward off a host of negativity from plagues and diseases to sorrow, lust and evil thoughts. In this tradition, ruby is a mineral which counteracts the "evil eye," and returns to the sender all negative thoughts, looks and feelings. Ruby is the July birthstone. Ruby is listed in the Spiritual Apothecary in the Etheric Sealer division. *(1st through 4th chakras)*

RUTILE (Red Crystalline): Rutile is a titanium dioxide forming in short, flat, red transparent and striated crystals. Rutile increases the operating efficiency of machines, adding higher levels of energy. This characteristic applies to humans as well, and is a powerful aid to degenerative issues of burnout and old age. Rutile, in its solitary form, is good for counteracting stress-related burn out; wear only in a relaxed state of use. Rutile may scorch sensitive nerve endings if used during times of activity, because the linings are too taxed to carry the higher level of current provided. This is particularly true for people with slow metabolisms, and should not be a problem for constant wear for highly active or hyperactive people. Retire to and relax in a comfortable setting, and use during a nap or meditation. People with attention deficit hyperactivity disorder may find that rutilated quartz acts much like ritalin. *(1st chakra)*

SAGEBRUSH (Petrified): See Chalcedony.

SAGENITE: Sagenite is characterized by fibrous fans of usually yellow stalks radiating in half-moons in a jasper-agate mix. It is a vision stone, as its name implies. Those who use it can emphasize their discernment abilities without being crass or authoritarian. This stone aids the tactless in seeing the impact of their words and behavior. It is a "lesson" stone; often useful to children to calm them so they can be spoken to, and they will remember and be influenced by what is said to them regarding their conduct or behavior. It is not a replacement for words well spoken. It allows for intensity, honesty, compassion and dignity in communication. *(6th chakra)*

SANIDINE: Sanidine is a potassium-sodium feldspar and is classified in the feldspar group. Having a disordered aluminum-silicon arrangement, sanidine is not structurally organized enough to exhibit high spiritual qualities. It is useful environmentally to reduce electromagnetic stress. It addresses the emotional center in similar fashion to moonstone, however accessing sufficient quantities may be a challenge. See Azulicite.

SAPPHIRE: Sapphire is the common name for corundum (aluminum oxide) which is colorless, gray, yellow, blue, green or purple. The asterism of corundum is caused by crystallographic inclusions of rutile. Sapphire is a profound aid to centering, and can be useful for inner ear disturbances which cause vertigo as well as cases of motion and sea sickness. It is also an aid to nausea caused by altitude and bad odors, and can be useful to undertakers, coroners, hospital and convalescent care workers, pregnant women, mountain climbers and hot air balloonists. From ancient times, sapphire was considered to counteract the evil eye as well as the plague. It carried the ability to ward off negativity of the spirit which led to hopelessness, so it became a favorite to dispel poverty, injustice and ill treatment by those in power. Sapphire is the September birthstone. Sapphire is listed in the Spiritual Apothecary in the Etheric Conditioners category. *(6th chakra)*

Blue - Blue sapphire amplifies intuitive ability, and is the prime mineral embodiment of transcendent wisdom. Light blue is an aid in meditation or prayer as it provides mental calming. Dark blue sapphire is considered a direct conduit to infinity, eternity and dreams. It is the emblem of detached spirituality, purity and truth. Blue and white sapphires evoke the energy of Our Lady. Blue sapphire retains the energy of the geomantic, or positive symbol, dragon. Blue sapphire represents air energy in the corundum family. *(6th chakra)*

Gray - Gray sapphire represents the earth element in the corundum family. Gray sapphire provides an excellent mineral connection for humans to whales, as gray sapphire resonates to whale medicine. It aids the process of sound vibration being translated to visual image as in echolocation. It represents the mid-point or Bindu point of consciousness. Gray sapphire bridges a strong connection between earth and self which deepens awareness of the divine heart beat of the Mother, which was heard by all in the silence before language. Gray sapphire aids recall of akashic records of collective consciousness, and can be called the "record keeper." The etheric conditioning provided by gray sapphire is greatly enhanced by toning or drumming during healing sessions. This increases vibrations to the etheric body which assists with DNA activation to access the song of self and one's true frequency. *(1st through 7th and 13th chakra)*

Green - Green sapphires aid the digestive system, helping with assimilation of food on the physical level. Green sapphire is the highest embodiment of hope and the turtle of life. Green sapphire represents water and feminine energies in the corundum family. It the emblem of salvation, safety and shelter. Green sapphire can be a symbol of dragon's blood as well as the hand of fate. *(4th chakra)*

Purple - Purple sapphire holds the story and mystery of reincarnation, rebirth and transformation both individually and collectively. It represents the hour of Christ's passion and transition from earthly incarnation to heavenly spirit. This is a stone of moderation in all things, of bringing the self into the Middle Way, of balancing experience of life with attainment of spiritual light. It is a stone of tantric power. In combination with a boron mineral, particularly danburite, purple sapphire holds space for psychological balance during the ritual of shamanic death. This is a combination of complete surrender which brings formidable power to the wise adept. Purple sapphire resonates to bat energy and the image of the hanged man in Norse mythology. *(7th and 9th chakras)*

Yellow - Yellow sapphire aids the processing and understanding of incoming psychic awareness on the cognitive level. Yellow sapphire can be considered the divine power source for royalty as well as immortality. *(3rd chakra)*

SARDONYX: Sardonyx is a chalcedony form of agate which is a banded combination of onyx and sard that is usually colored white and red-brown. This stone helps alleviate depression and lack of confidence. Since the impact is on the upper lungs, thyroid gland, and upper part of the nervous system, balanced behavior and increased happiness are the key notes. Its dark and light layers provide visual stimulation for polarity balance. Sardonyx is the August birthstone. Sardonyx is a Healer in the Spiritual Apothecary. *(5th chakra)*

SCAPOLITE: Scapolite is a group name for complex silicates which contain sodium, calcium, potassium, silicon, aluminum, chlorine, sulfates and carbonates. It is shaped like tourmaline in round to rectangular crystals with striations on its edges. Members of this family are useful for problems in the transverse colon, the intestines, bladder, colon, kidneys, liver, pancreas, bile duct, spleen and rectum. It is a powerful body cleanser which helps to release fat. It is antiseptic. It helps to form amino acids and metabolize proteins. It helps skin and fingernails. This combination helps to counteract obesity, high blood pressure, and carbon dioxide build up in tissues. Scapolite is helpful to counteract symptoms of old age, arthritis, bone loss and acidity. Colors can range from white to blue, pink, purple and gray.

Clear/Yellow - Scapolite is useful in helping individuals to break problems down into manageable chunks so a time plan of action can be visualized. It is especially tuned to water signs of the zodiac, or for people with several water placements in their chart. It is useful for dowsing, especially for water. Other colors of scapolite resonate differently.

Pink - Is an excellent mineral for children or for healing the child within. It is very gentle and soothing. *(4th chakra)*

Purple - "The Forester's stone." Purple scapolite is shaped in thick, tabular, and rectangular crystals. When Forester is capitalized, this indicates a member of a major benevolent and fraternal order, such as the White Brotherhood. This mineral was worn by St. Germaine as it resonates to the life force energy of the violet ray. It supports energy work in this area, such as Reiki and alchemical processes. *(6th through 12th chakras)*

SCARABS: These are carvings of any mineral into the sacred beetle shape of Egyptian origin. To them the hieroglyph, or symbol, of the scarab meant eternal life, or literally, to revolve or return. It symbolized resurrection, immortality, eternal renewal, and the cycle of the sun, the god who returns. Scarabs were seen to roll balls of dung in which were laid embryos. Scarabs were drawn holding the ball of the sun aloft or rolling the ball of resurrection from which new life emerged. Scarabs were placed on the heart of the deceased, because the Egyptians believed the conscience and morality resided in the heart, which would ultimately judge the conduct of the deceased. G. Posener in *A Dictionary of Egyptian Civilization* quotes words on an Egyptian coffin in the Vienna Museum, "A man's heart is his own god."

SCHEELITE: Scheelite is a tungstate of calcium. A useful quality of scheelite is its ability to infiltrate a lie or obfuscation. It can be used to determine if someone is dishonest with you, and to delve into personal issues of denial and falsehood. Scheelite is brave like a scout and will provide courage to those who are intimidated by a falsehood. It activates the third and second chakras to empower the gut to not only sense danger, but to do something about it. It is excellent for interrogation, interview and detective work. It is helpful to those in a disciplinary job such as school officials, guidance counselors, drug and alcohol counselors, and probation officers. It embodies mouse medicine. *(2nd and 3rd chakras)*

SCHIST: Schist is an overall term for several varieties of metamorphic rock. The varieties are mica, chlorite, prasinite, talc, glaucophane, and calc schist. The essential components of mica schist are quartz and mica, and chlorite schist must contain chlorite. These first two schists were often used for roofing material. Prasinite must contain chlorite, actinolite, albite and epidote. The essential component of talc schist is talc, and it is the principle source of talc industrially. Glaucophane must contain glaucophane and lawsonite. Calc schist must contain calcite, mica, chlorite and quartz. Schist is an excellent cleanser, toner and balancer for the fingers and meridians leading off the fingers.

Depending on what is combined with schist, specific attributes can be promoted. The color of the stone or conglomerate material will most clearly define its application. For example, silver mica schist with garnet in it creates a body balancing circuit through the index fingers. This combination is particularly tuned to creating unity between the etheric and auric bodies, a wonderful experience for those unaccustomed to spiritual awareness with a grounded and "whole" feeling. It is a useful tool for those who have only intermittently and with intense concentration reached spiritual connection, because it eases (like lubrication) the ability to connect. The natural harmony of chosen mineralization again shows how specific connective attributes are especially attuned to the needs of the many in their quest for spiritual enlightenment. Gold schist held at the throat chakra by the right thumb creates a body circuit which allows the gold healing ray to be breathed into the body. The throat is often experienced as an emitter because of the vocal cords, but it is the primary chakra conduit of physical, as well as etheric, well-being because of ingestion and inhalation. The ability to breath in healing essence can be accomplished with specific mineral/finger combinations, and useful to study. For the body system, clear quartz and gold are the strongest overall healers, protectors, and balancers, but schist is an affordable option which works optimally. Schist is categorized in the Healer division of the Spiritual Apothecary. *(Fingers and meridians)*

SCOLECITE: Scolecite is a hydrated calcium aluminum silicate classified in the zeolite group. It physically influences lupus erythmatosus, and in reflexology, the fleshy area of the hand between the thumb and index finger, and therefore those areas of the body impacted by performing reflexology in this area. It also alleviates feelings of guilt and fear of not being wanted, not belonging, or not being good enough. *(7th chakra)*

SELENITE: Selenite is the common name associated with clear and colorless gypsum which is calcium sulfate. Because selenite (and gypsum) can form quickly in the right conditions, it is a mineral which is in touch with current events on earth. It bears a current seeking attribute which aligns its properties to consciousness evolution, and it is an important mineral to assist futurist and prediction accuracy as well as to stimulate staying in present time. It is a mineral which brings in the Now. Selenite is listed in the Auric Openers division of the Spiritual Apothecary.

Black Satin - Black satin selenite can be used in daily prayers. It helps an individual to focus readily on and maintain an orientation to the Mayan calendar. It provides a strong sense of spiritual continuity over time. A very complete mineral combo. *(1st and 7 chakras)*

White/Clear - "Halo Rock." This clear form of selenite is a mental and spiritual expander. It is a magical and powerful tool with particular benefit to practitioners of healing, meditation, and alchemical and psychic arts. Many of the clusters have absorbed mysteries which can be unlocked and revealed to the sensitive or psychic individual.

Selenite reduces swelling of mucus membranes. It acts as a tension reliever and balancer of the body's electromagnetic field. Unparalleled as a psychic state inducer and supporter, selenite opens the crown chakra. It is recommended to be used with grounding and protecting stones. The arrowhead, desert rose, hourglass, floret, and fishtail shapes show its variety of powers of manifestation. Calcium is a feminine element and sulfur is masculine. Sulfur was thought alchemically to be the elemental representative of spirit fire. The combination of these two elements in clear crystals provides an unequaled energetic pulse for the physical body to receive spirit frequencies. It is not recommended to be placed near sleeping quarters without sufficient grounding stones for balance as it will likely disrupt sleep and may cause sleeplessness or insomnia. For those who already suffer from insomnia, selenite may counteract this problem. It is an excellent choice for addressing broken bones, bone cancer and infection, and ulcers. Calcium is the most abundant element in the body, and must be renewed every six years. It is felt that psychic and intuitive work leaches calcium from the bones to support cellular functions working hard to move light frequencies through the nervous system being brought in by sulfur. Selenite alleviates psychic strain while it simultaneously acts as a bridge to the bones to increase psychic confidence, sensitivity and openness. See Gypsum. *(7th chakra)*

SEPTARIAN NODULE: A septarian nodule is a sand concretion of a mudball where yellow calcite crystallized in cracks, which can be narrow and minuscule at first, then open into large cavities. They are found in Utah.

Red - Red septarian nodule heals the wall of the body cavity. It also opens and closes the body wall for psychic surgery. Study the structure of red septarian nodule, and you will understand how these attributes are realized. Work with this mineral to strengthen psychic healing abilities. *(1st chakra)*

Yellow - By far, this is the more common variety. Yellow septarian nodule is an excellent kidney cleanser. It is a good mineral for depressed people as it enables personal insight to drive changes toward healing. It is a good mineral for children's rooms because it is supportive and counteracts nightmares. If a child reacts negatively to it, psychic work is necessary to clear the room of unwanted energies. *(2nd and 3rd chakras)*

SERAFINITE: This is a massive, acicular chrome chlorite from Russia. Chlorite is a group name for minerals with micaceous tendencies. Chlorite is often included in quartz and is green. Massive serafinite grows in green fans with blue fire iridescence. It is uncommonly beautiful. Called "the Medium Stone," serafinite reinforces oral communication abilities of spiritualistic mediums. It strengthens the connection to the parties involved in communication and allows the medium, as intermediary, to remain more present to self without operating in a trance state. Serafinite is not recommended for use during sleep, as it allows disintegration of life cord bonding to physical self. This occurs during sleep, because mediumistic powers tend to flow from physical bonding to the higher self where physical essence may choose to remain. Serafinite holds huge amounts of healing vibration, and is useful to all healing applications. It can be used to amplify the effect of other minerals, but it provides sufficient energy on its own to bring in wholistic awareness. It serves diagnosis. *(Upper chakras)*

SERANDITE: Serandite is a hydrous sodium manganese calcium silicate shaped in orange, tabular and rectangular shaped crystals from Mt. St. Hilaire, Quebec. Serandite promotes autoreticulation which is the automatic formation of networks. It can strengthen evolutionary change of the human brain toward being capable of harnessing higher frequencies of light (kundalini). It fosters the production of phagocytic cells which engulf foreign bodies harmful to the body. It is a superior cleansing mineral, and it regenerates and supports the kidneys when combined with microcline or orthoclase minerals such as moonstone and labradorite. It nurtures the mesenchyme which produces connective tissue, blood, lymphatics, bone and cartilage. Serandite is a Healer in the Spiritual Apothecary. *(2nd chakra)*

SERPENTINE: Serpentine is a group name for hydrous magnesium iron silicate minerals as well as a type of regional metamorphic rock. It comes in various shades of lime to darker green, often with a greasy texture. Antigorite is the mineral name associated with the dark green type of serpentine which is a powerful healing stone. Serpentine's color associates with the heart and lungs, but it is also important to the liver and brain. It incinerates toxins, and activates the absorption of nutrients and oxygen. It helps fever to cleanse the body of microbes. There is a tendency for psychic opening as cleansed core body functions enhance the individual's sense of security, and relaxes one's overall nature. It is highly applicable to the nervous system, and alleviates chronic fatigue, irritability, and mental illness and retardation. It is helpful to reduce muscular fatigue and cramping. The apple a day concept, to keep the doctor away, can be fulfilled physically by combining serpentine with calcite and turquoise. For all issues involving the nerves, whether stress related or due to illness, combine serpentine with turquoise, variscite, apatite or vivianite. The state stone of California is serpentine. Serpentine is located in the Healer category of the Spiritual Apothecary. *(4th chakra)*

SHATTUCKITE (Blue Chrysocolla): Shattuckite is a hydrous copper silicate differing from dioptase only in the number of atoms and crystal shape. Shattuckite is orthorhombic while dioptase is trigonal. Shattuckite can be confused with azurite, because it associates with malachite and chrysocolla where it is found. Shattuckite is an earth tone mineral full of vibrancy and power. The many meanings of the word "tone" are active in this mineral, allowing it to enhance sound healing, physical health through toning, chanting and singing, understanding of communications, and color therapy. It is excellent for understanding the personality nuances of others. Shattuckite is an excellent transmitter of earth energies and qualities, and can be called a "space ambassador," for other life forms in the universe can receive much intuitive understanding of Earth from this mineral. Also see Kinoite. *(6th chakra)*

SIDERITE: Siderite is an iron carbonate forming two series with magnesite and rhodochrosite and is classified in the calcite group. It is beige colored, and can crystallize in blades or discs. Finding or achieving being centered often requires loosening the solar plexus. Siderite facilitates healing through breath and meditative work to relax the solar plexus. It helps the individual to trust self-determination and to breathe in desired positive qualities. *(3rd chakra)*

SILICON: Silicon is always found in nature combined with oxygen. Quartz is the most common and simplest silicate. To be utilized in technology, quartz sand is heated to de-bond the silicon and oxygen atoms. When it cools, the surface is covered with a pattern reminiscent of meteorites which is attractive to some jewelers. In its purest form, silicon (silvery gray) is a mineral element always reaching for the future. It is an ally for people working with futurist themes, such as commodities, stocks, city planning, space exploration and prediction. Its reflective nature offers protection similar to hematite. *(1st and 7th chakras)*

SILLIMANITE: Sillimanite is an aluminum silicate also called fibrolite. It is trimorphic with andalusite and kyanite. It is usually opaque, and forms in a fibrous aggregate (fibrolite) or frequently as an acicular crystal inclusion in other minerals, such as quartz and feldspar. It is good for balancing retrograde positions of planets in the solar system, and to use in meditation when the moon is void-of-course. It enhances hearing or clairaudient as well as clairvoyant capacities. In the way that a picture is worth a thousand words, sillimanite stimulates the ability to formulize images from incoming psychic frequencies. The best way to describe this is to compare this essence to how cetaceans (dolphins and whales) echolocate and are able to read their environment from sonar pulses bouncing back to them from objects. Ultrasound technology was developed from our study of cetacean echolocation, and these aluminum silicate minerals potentiate this psychic ability in humans. Sillimanite is the State mineral of Delaware. See Andalusite and Kyanite. *(7th chakra)*

SILVER: Silver is an element listed just above gold in the Periodic Table of the Elements. Argentum, the Latin word for silver, derives from the Sanskrit denoting "shining white." Silver is a general disinfectant and cleanser as it works on the entire physical and subtle body systems. It is an emotional and mental stress releaser due to its purity and conductivity combination. The five highest chakras are activated when this element is worn. Silver resonates to lunar, water, receptive and feminine energies. It is a reflector and deflector. It harnesses the ability to see clearly and discern, because the moon can be gazed upon directly without hurting the eyes (as the sun will do). It is the metal of shamanic healing or night eagle (owl) energy, and the metal which best transmits long distance healing. Many cultures revere silver as a symbol of purity and purification. It also denotes clear conscience and consciousness. In some Russian cultures, people were healed of epilepsy, because they relaxed as they watched a healer work silver into various shapes. Silver is the State mineral of Nevada. It is listed in the Spiritual Apothecary in the Healers division. *(8th to 12th chakras)*

SMITHSONITE: Smithsonite is a zinc carbonate listed in the calcite group. When it grows botryoidally, it can be polished, but it usually forms in a thin layer. All forms of smithsonite serve to improve personal and spiritual interrelationships.

Blue/Green - This is by far the most common color. When infused with copper, smithsonite is blue or green. In the turquoise color, its key ability is to connect and align the emotional and astral bodies through the throat. This ability enhances bodily electromagnetic balance helping with all manner of neurological disorders. It also strengthens the immune system to thwart allergies. *(4th and 5th chakras)*

Brown - Smithsonite is brown when infused with iron hydroxide. *(1st chakra)*

Pink - Smithsonite turns pink when infused with manganese. In pink and green, smithsonite calms and releases negativity from the heart area by helping an individual to cope with the opinions, power plays, dramas and egotism of others. It provides freedom from identifying with the tribal forces of family and community, so an individual can activate life purpose which may have been thwarted by the controlling efforts of others. Pink and green are unbeatable as a pairing to build the immune system. Pink smithsonite by itself protects the immune system from emotional attack especially if an individual is vulnerable to the energies of others. If an individual takes on the illnesses of others, pink smithsonite establishes a boundary layer of immunity. *(4th chakra)*

Violet - Smithsonite turns purple when infused with cobalt. The violet color soothes and regenerates the sixth chakra, and it activates high spiritual, creative and mental awareness. *(7th chakra)*

Yellow - When infused with cadmium, smithsonite is yellow. *(3rd chakra)*

SOAPSTONE: Soapstone is the common name for a form of talc (hydrous magnesium silicate) called steatite which is generally easy to carve. It bears a variety of colors, often gray, depending on elemental impurities, often green, and it feels greasy to the touch. It is a poor conductor of heat so it has been used for spark plug insulators and giant high voltage insulators. Soapstone is an essential mineral for the home and other daily use environments, because it harmonizes so well with mundane activities. It is an excellent "workhorse." It orients other rocks and minerals to place, providing a solid foundation for the minerals. Specimens can be placed on the soapstone as a general "home," and since soapstone balances and counteracts static, minerals are maintained in optimum working order all the time. The thickness of soapstone does not matter, nor the color. Soapstone can be cleansed with smudge, incense, flame, and sun applications.

SODALITE: Sodalite is a sodium aluminum chlorosilicate classified in the sodalite group with lazurite, a constituent of lapis lazuli. It is colored a deep blue with white calcite veins in most pieces. Sodalite possesses a strong feminine vibration. If lapis lazuli feels caustic to the nervous system, try sodalite. Its high blue color shows its spiritual signature trait. It is an affordable replacement for lapis lazuli in healing and spiritual growth work and in providing activation to the third eye area. Sodalite is excellent for cleansing the body and blood, reducing fat as a byproduct of this cleansing. Metabolic balance and cleansing of the glandular and lymphatic systems is an important affect of this mineral. Sodalite stimulates the adrenals and enzyme production, and assists the liver, pancreas, spleen and bladder. It can be thought of as the banana, pineapple or celery of the mineral kingdom as these plants contain appreciable quantities of chlorine which helps third chakra organs as sodalite does. (3rd through 6th chakras)

SPECTRALITE (from Finland): Spectralite is a form of labradorite which is a plagioclase feldspar. It is an infuser stone. The schiller colors found in spectralite exhibit the shimmering quality of aurora borealis, a phenomenon of the high northern latitudes. These Finnish rocks are infused by the awe-inspiring essence of light crystallized in ice in the sky. When the aurora borealis occurs, it is ephemeral and shadowy as well as dramatic and flaring in intensity. The ethereal essence of this sky show is retained in spectralite, and the delight of light movement within the rock's structure calls forth the "yes" response. Its infuser nature inspires the higher self to enter form. This mineral can be worn by all astrological elements. Air signs are dignified by being the canvas upon which beauty is displayed. As crystallized water is the necessary paint on canvas of the aurora borealis, the invaluable role of water signs in the scheme of human endeavor is shown to others. Earth signs are validated for their ability to sustain the ephemeral passage of beauty for others to see longer than a moment. Fire signs are shown how to modulate and focus their vibrancy so their beauty is not consumed and burnt out too quickly. Through the process of using this stone, one realizes that beauty is a natural phenomenon which flares from within to be showcased at times on an expansive canvas upon which miraculous and magical powers can be known and seen by self and others. *(All chakras)*

SPHALERITE: Sphalerite is a zinc iron sulfide classified in the sphalerite group. Brown to yellow in color, with some translucence, sphalerite can appear greasy. In crystalline form, sphalerite is known as an "Earthsong" mineral embodying the spirit of Earth to move and make sound, such as volcanic eruptions, earthquakes, and shifts of axes. When the Earth moves, mountains sing deep, throaty choruses, the bass notes of Earth's orchestra and chorale. Sphalerite is an energetic complement of fluorite, providing infusions of well-being after release of pain and negativity by fluorite. Sphalerite incinerates toxins and begins their movement out of the body, and then cauterizes the edges of energetic wounds. Opaque forms of sphalerite are useful to help the healing of ear and hearing problems. It is an excellent mineral for toning and teachers of voice. *(1st and 5th chakras)*

SPHAEROCOBALTITE: Sphaerocobaltite is a cobalt carbonate listed in the calcite group. Colored pink to magenta, it forms as a coating, and sometimes it crystallizes. It is found most often in Central Africa. Sphaerocobaltite is an excellent heart chakra and solar plexus connector. When someone says, "My heart isn't in it," it is often because the solar plexus is blocked and the heart cannot discern intention nor receive inspiration, therefore commitment will not follow. Sphaerocobaltite clears the solar plexus to operate as the intake center for the inspiration for life, and allows the heart to feel supported and energized by flow from the solar plexus. Sphaerocobaltite is known as the "weaver's stone" for many visions woven in cloth are the result of clear and healthy connection between the solar plexus and heart chakra. It is related to spider, especially black widow, because she bears the characteristic red mark of this mineral. *(3rd and 4th chakras)*

SPHENE (Titanite): Sphene is the common, old timer's name for titanite which is a calcium titanium silicate. It is colored yellow-green to dark brown in clusters of short, tabular crystals. Working with this mineral aligns the spiritual, emotional, etheric, and mental bodies. The entire system of nodes and meridians is stimulated, and general health is thereby improved. It has applications to shape shifting. Sphene is one of several minerals which softens human energies and allows bonding to occur with other animals. Since it is not readily available, it is usually reserved for high ceremonial uses. *(All chakras)*

SPINEL: Spinel is a magnesium aluminum oxide classified in the spinel group. It occurs as octahedrons in twins or rounds on matrix. Spinel can be white, blue, green, pink, brown, red or black. It is a powerful nervous system vitalizer. This mineral helps synchronize the physical, emotional and etheric bodies. All forms of spinel in one way or another assist with seeing something which is difficult emotionally or mentally and in visualizing something which is difficult to attain or manifest. Depending on the color of spinel, chakra activation determines the character of what is able to be seen. *(Chakra varies with color)*

Blue - The blue variety is a detoxifying agent for poisons, and allows an individual to see the true character of those in power and positions of authority. *(5th and 6th chakras)*

Brown/Black - Brown and black spinel can be useful to maintain balance for those who see horrible things as the nature of their business or jobs. This includes detectives, police personnel, coroners, undertakers, etc. *(1st chakra)*

Green - Green spinel adjusts the pancreas to assist with diabetes and alcoholism recovery, and increases the ability to see self clearly and with compassion. *(4th chakra)*

Pink - Pink spinel is an excellent mineral for those emerging from the closet, revealing their true selves, and wishing to be seen for who they are. This can be true for homosexuality, psychics, people in multiple relationships, or whatever is threatening to the dominant status quo. Pink spinel promotes seeing with compassion what is really there, and to acknowledge the importance of being authentic rather than seeming to be something else. *(4th chakra)*

Red - Red spinel is all about doing what one envisions, bringing visions into form and manifestation. It brings in the energy of antelope. *(1st chakra)*

Ruby Spinel in Biotite Mica - This mineral forms red, pea-size spots in a black matrix. This rich conglomerate grounds the entire system and amplifies the power of the root chakra. It enhances courage, stimulates drive and action, and provides foundation and focus. For individuals lacking a firm "footing" energetically, it promotes connection to the here and now, to place, and internal balance. It provides an excellent base of support, a firm rooting. *(1st chakra)*

White - As a team, red and white spinel work on nervous system renewal, reducing inflammation of nerve fibers, and increasing the ability to "be in the world but not of the world." *(7th chakra and above)*

SPODUMENE: Spodumene is a lithium aluminum silicate with vertical striations (like tourmaline) which comes in white, yellow, green (hiddenite), gray, and pink to lavender (known as kunzite). It is a mineral family with several color expressions. Spodumene has internally a prismatic structure, and on the outside bears a refined columnar shape. Spodumene elevates consciousness. Lithium-rich, spodumene is uniquely qualified to transmit the lavender ray directly into the nervous system and physical life. Lithium is the elemental link between light (energy) and body (matter). Lithium ranks third after hydrogen and helium as the lightest element. It is not natural to earth, and has arrived on earth in minute amounts compared to other elements only through cosmic radiation. Lithium turns minerals pink, lavender and purple. All colors of spodumene are found in the Healers division of the Spiritual Apothecary under the name Kunzite. *(Chakra varies with color)*

Hiddenite (Green Spodumene) - Green spodumene is a kunzite cousin. Hiddenite is a vision quest mineral for those seeking the place of one heart. It should be worn at the heart chakra, and should be the only mineral worn on the body during a young person's first vision quest, though other mineral tools may be used and carried to the site. Hiddenite is a full spectrum mineral with vibratory influence from the fourth chakra above the crown to the foot chakras. It is an excellent visionary tool. *(4th chakra especially and all chakras)*

Kunzite (Pink Spodumene) - Kunzite is a regenerative agent for the cardiovascular system. It opens heart/love in a balanced way, connects to highest life and love forces, and promotes soul unification. It provides stress relief from emotional traumas with wearing and meditation. It is the primary mineral conductor of Peace to Earth. When turned sideways, the pink deepens. The highest grade kunzite is nearly purple at this angle. Kunzite is found in the Spiritual Apothecary in the Healers division. This form of kunzite is highly recommended for healing layouts as it reinforces and strengthens the transmission of life force energy. Kunzite is recommended for those wishing to make or maintain contact with the deceased. If the death was accidental or resulted from suicide, kunzite is not recommended for a period of time until the spirit is received on the other side and brought to a state of awareness and acceptance. *(4th chakra and up)*

STAUROLITE (Fairy Cross Stone): Staurolite is a chemically complex hydrous oxide of aluminum, iron, magnesium and zinc. In black or red-brown, its characterized most often by cruciform twins, though it can occur as single crystals It is colored a light tan to dark brown, and forms in flat, pencil-like tabbie crystals. The 90 degree Greek Cross and the 60 degree St. Andrew's Cross shapes of staurolite are most prized for luck and spiritual values. Staurolite is an excellent physical protection mineral which enhances the descent of the total self-light into the physical plane. It is exceptionally useful for disturbed children who can easily fall prey to unseen negativity when distress is in their lives. Staurolite is an inducer of higher self-perception, supporting an individual to remain in personal truth, thereby creating a reality and sensation of courage as well as the ability to overcome denial. Whenever battle must be done with the forces of dark, staurolite is a well-honed companion for the task. The 90 degree form of staurolite is the stronger form for personal protection. The 60 degree form is more conducive to higher self recognition and positive visualization. Both forms possess aspects of the other, one choosing to orient more to the other use than the other. Staurolite reduces the effects of stress, soothes the nervous system, builds and detoxifies the blood, and supports the immune system. Staurolite is the state mineral of Georgia. *(1st chakra)*

STEWARTSONITE, TACHYLYTE and TITANIUM: Dark blue and metallic, this combination stimulates precognition as well as analytic reasoning abilities. Excellent for use during research, investigation and psychic activity. *(6th chakra)*

STIBNITE: Stibnite is a simple antimony sulfide which is soft and heavy. It is colored steely black/gray with an iridescent coating on its normally acicular or flat, striated crystals. China produces premium specimens.

Stibnite draws spiritual perceptions and awareness deep into the physical body, allowing the individual to look at the internal system to determine if disease is present. Therefore, it is a useful diagnostic tool. It applies well to hypnosis and meditation, and can alleviate fears or difficulties of skittish meditators. It is an excellent centering stone. *(All chakras)*

STILBITE: Stilbite is a hydrated potassium aluminum silicate which can be either calcium or sodium rich. It is classified in the zeolite group. It is usually peach colored, and forms in flat, bow tie crystals with a "sheavy" growth. It is often found associated with apophyllite. Stilbite is categorized in the Spiritual Apothecary in the Healer division.

Orange - This type of stilbite is useful for alleviating physical symptoms of menopause, because it regulates body fluid balance and relaxes muscle fibers. It supports flushing of waste products from the kidneys and liver. It balances issues of tissue dehydration, and helps to control hot flashes and mood fluctuations of anger and frustration. It counteracts osteoporosis. It is invaluable as a mineral which honors the crone-phase of female power. Its gentle purpose sympathizes with the changing strength of physical prowess into spiritual/psychic prowess. It can be used by men for transition of male power and geriatric issues. *(2nd chakra)*

Peach - Especially in the peach tone, stilbite is a special mineral used for bridging the arc into nonphysical reality, particularly when vacating the body at death. It is harmonic with rose quartz and aurichalcite, often creating the necessary heart chakra connection and healing which allows the individual to accept and yield to the higher self guidance of choosing to vacate the body. The body is often healed by this combination, thus making clear that spinal or mental afflictions were causing the physical descent. Stilbite is a powerful relayer of the loving allowance which creates the ability to accept one's condition and move beyond. *(Upper chakras)*

STISHOVITE: Stishovite has the same chemical formula as quartz, but its structure is tetragonal, allowing the formula to express itself as metallic rather than clear. It is categorized in the rutile group. Stishovite strengthens operational predictability of machines and humans, and sets a standard of consistency which is transmitted to machines and technological systems.

STICHTITE: Purple or lilac (stichtite) and green (serpentine) are the Pleiadean colors. We found some stichtite where it was first discovered on the west side of Tasmania, near Zeehan. Stichtite is a hydrous magnesium chromium carbonate listed in the hydrotalcite group. It is dimorphous with barbertonite. Stichtite enhances connection to the Pleiades. Its soft soapstone texture enhances "melding" with its color rays, thereby increasing physical resonance and sensations between the fourth and seventh chakras. Stichtite moves heart healing into consciousness shift. It is an excellent mineral for counselors. Other forms of purple and green minerals will not provide exactly the same vibrations, however fluorite can be found which combines these colors. *(4th through crown chakras)*

STROMATOLITE JASPER: A bio-feedback mineral which enhances mental release or self-hypnosis so that the physical autonomic processes can be felt, gauged and worked with. Stromatolites are laminated sedimentary fossils formed by layers of blue-green algae.

STROMBOLITE (Spurrite): Strombolite is a common name for spurrite which is a rare limestone occurrence of calcium silicate carbonate. It is lavender with white seams of calcite. Strombolite is appropriately called the "storyteller stone" metaphysically, because it links silicon and carbon with calcium. It is the holder of the story of life through the kingdoms of plants and animals, both terrestrial and aquatic. It reduces inhibition in public oratory, allowing the full power of the speaker to be expressed, because its authenticity is apparent. It smoothes diction, and helps words to flow easily, with pleasure and skill. It is naturally a support mineral for archeologists, paleontologists, and anthropologists to name a few. *(5th and 6th chakras)*

STURMANITE: Sturmanite is a hydrated calcium iron aluminum manganese sulfate borohydroxide listed in the ettringite group. It forms in yellow, bubbly crystal masses. Good specimens come from Tsumeb, Namibia, Africa. Sturmanite is an excellent mineral to facilitate dissolving tubal blockages such as in the Fallopian tubes or urethra. It needs to be worn at the throat for best effect. *(2nd and 3rd chakras)*

SUGILITE: Sugilite is a potassium sodium iron manganese aluminum lithium silicate classified in the osumilite group. Found in Africa, sugilite varies from a deep purple to a violet color and can also express itself as dark blue. It can be mixed with hematite. Its supremely high properties are well utilized in the Spiritual Apothecary in the Healer division.

Blue (Richterite) - Blue sugilite can be exemplified by the term "sfumato" from Webster's which refers to "the definition of form without abrupt outline by the delicate blending of one tone into another." Sugilite, in general, does an excellent job of melding elements for the highest good of humanity. Humans have been endowed with tremendous potential for wisdom since before the time of Shelomoh or King Solomon, who was noted for his wisdom. It is said of humans that they know not the whole of their purpose upon the Earth, because humans succumb to their frustrations in their urge and zeal to know. A simple truth that humans appear to learn a new in each generation is that wisdom comes to those souls not needing to know in order to be. Separation is caused as much by the need to be recognized as learned and right as it is by the exercise of judgment and bias. More wise people have been murdered on this planet than any other. Is this not an indication of the need to possess wisdom, and if not, the need to destroy it? The story of the Jewish people is instructive in this stone, yet the twentieth century is filled with many examples of intelligentsia genocide.

The intellect (blue sugilite) must be wedded with the highest urge toward spirituality (purple sugilite) in order for man to reveal the most compassionate tendencies of his mind, which is the union of heart and brain. Sugilite of blue and purple mix is the stone of rest in one's life, signaling the time of quiet devotion, contemplation, communion and peace. To honor the gifts of this stone, it can be consecrated on the old Shabbat (Saturday Sabbath) as a facilitator of one's desire to let go of the need to conquer, control, build or amass. We are all one as we blend our tones together without pointing to the outlines of our skin or culture, sex or faith, intelligence or wealth. Wisdom has always been the gifted potential of humans, but the apple made wisdom seem as if it could be possessed, as well as separate, outside and dangerous. Much has been wasted and lost. This stone of remembrance awakens the heart/trust knowledge that another person's wisdom may be the key to unlocking your own. Honor the ways of your enemies and you will have none. Honor your way of knowing and you shall be free. Come together as free spirits and you shall inherit your purpose. *(5th through 7th chakras)*

Purple (Luvulite, Royal Azel) - First discovered in 1975, it wasn't until the end of 1980 that Lou Koekemoer of South Africa was able to prove that the deep purple material he suspected to be special was manganoan sugilite. The only supreme purple color in the world comes from a pocket in the Hotazel area of South Africa. Professor Kenici Sugi first discovered the mineral in 1944 in an aegerine syenite on Iwagi Inlet, Ehime Prefecture, Southeast Japan. His original assay was correctly identified by Dr. Hisoshi Matsunaga several years later. Royal Azel is the name of crystalline sugilite which can be faceted. Gem sugilite, when faceted, changes color from reddish-purple in incandescent light to bluish-purple under daylight or fluorescent light. Most material is massive. Sugilite is a silicate of potassium, sodium, iron, lithium, manganese and aluminum. It is classified in the osumilite group.

Sugilite is an excellent mineral to induce calm and balance to the brain, heart, muscles and nervous system. Sugilite helps to prevent constipation and to normalize digestion. The deep purple color creates balance for the body's uppermost glands (pineal and pituitary), and connects spirit forces with the nervous system. It emanates a loving, gentle tone of compassion and forgiveness. Its healing attributes rain on the physical and subtle bodies, helping with all modern stress-related diseases as well as HIV/AIDS and cancer. As stress causes a constant fight or flight reaction in the body, all body functions begin to deteriorate. Sugilite helps to counteract and reverse this trend. It is valuable to empaths, psychics and people with sensitive nervous systems, including children. Sugilite prevents energy vampirism or feeling consumed by the needs of others. It helps sensitives to cope in densely populated areas where pollution is not just industrial but psychological, emotional, and spiritual as well. It was discovered none too soon to assist with consciousness evolution. *(6th to 7th chakras)*

SULFUR: Sulfur is a native element. It is commonly obtained as a byproduct of the oil industry; however, it is also the byproduct of ancient bacteria. Its sunny color is the unlocking force to enhance physical vitality and beauty. Sulfur acts as a poultice to draw impurities from the body's lymphatic system, relieving infection, colds and sinus problems, and helping to build the immune system. It's a great infection fighter. Sulfur dioxide is especially toxic to molds and some bacteria. Hydrogen sulfide is a more poisonous gas than even cyanide gases. It destroys respiratory enzymes, and even small amounts cause headaches and nausea. Since insulin is a compound of sulfur, this mineral is excellent to place environmentally and to use in healings for diabetics. Sulfur drugs are utilized for many skin problems, because its action in the body drives wastes and toxins out through the internal excretory system thereby clearing the skin. It aids in hemoglobin and amino acid production and metabolism of amino acids. It is applicable to overall health, but it relates well to maintaining the health of all organs in the second and third chakra areas, including the reproductive and elimination system organs.

Alchemically, sulfur represents the spirit (see mercury, salt and cinnabar). Sulfur is thought to be released from its limitations by being combined with mercury into a physical, living understanding. Further alchemical processing of masculine sulfur and feminine mercury resulted in the production of gold, a perfect union. It is interesting to note that cinnabar, a mercury sulfide, was for the Chinese a medicine of immortality. They ingested mercury which led to early death. Sulfur represents the fire principle, and was characterized as semen in mineral form. Sulfur helps to strengthen connection and surrender to Divine will.

In a negative sense, brimstone (sulfur) rained down on Sodom, and was considered diabolical in the Bible. It became for the Christians a symbol of light turned into darkness, because pagans used it to cleanse the guilty. As is common in Christianity, many positive earlier beliefs and practices were turned negative. For Islam, red sulfur became a symbol of a man of integrity and the process of soul transmutation through ascetic practices. The Romans utilized sulfur to make matches, lamp wicks, explosives, bleaches and medicines, much as we do today.

Globally, a nation's industrial development is measured by its per capita consumption of sulfur, because it ranks consistently as the most widely produced industrial chemical in the world. Sulfur exists in more allotropic forms than any other element. This property results from the ability of sulfur atoms to combine with each other in long chains or in rings.

Environmentally, sulfur acts as a sterilizer and disinfectant. It can be thought of as the onion of minerals, because onions contain abundant amounts of sulfur as does garlic, horseradish, cabbage, avocado, pineapple, Jerusalem artichokes, and soy beans. Bitter vegetables contain the most sulfur. One reason soy beans cause allergic reactions is they dry out the bowel with too much sulfur. Sulfuric acid is used to remove water from organic compounds in a reaction that generates substantial heat, because it possesses a strong affinity for water. In fact, nature's analog for water, H_2O, is H_2S, hydrogen sulfide. Ancient mounds of sulfur are being mined today that are the result of primitive bacteria breaking down H_2S for hydrogen for photosynthesis, leaving sulfur as a sewage byproduct of decomposition. Sulfur is listed in the Spiritual Apothecary in the Etheric Cleanser category. *(3rd chakra and skin)*

SUNSTONE (Heliolite): Sunstone is a yellow orthoclase which can be colored all one tone or have a gold schiller area inside. Orthoclase is a potassium aluminum silicate. Specimens from Oregon can be orange with green schiller. In the yellow and darker shades, it projects light into the lower heart chakra area, causing facilitation of friendly will force which is harmonious, and meets little resistance. Sunstone is a valuable yellow mineral for cleansing and rebuilding the kidneys and liver, and can be an important aid to counteract cellular issues and degeneration. It aids proper digestion. It reduces acid conditions. Much like the Coriolis force, sunstone moves energy to the right around the body in ever-widening spirals, and this direction holds true for the northern hemisphere. The opposite is true for the southern hemisphere, where sunstone moves energy to the left around the body. This movement removes unwanted energies from the aura. While this stone is present, the aura is safer to open for healing work. Sunstone is the state gem of Oregon. Sunstone is an Auric Cleanser and Opener in the Spiritual Apothecary. *(3rd and 4th chakra)*

SUSTAMITE: Sustamite is a rich pink to purple stone, which looks like sugilite and is from Africa. Sustamite causes deep muscle relaxation and waves of relaxation which are conducive to massage, hypnotherapy, meditation, and prayer. It is like the plant, curare: a little goes long way. It aids in balancing electromagnetic impulses, brain waves, and heart rate. *(All chakras)*

TAGUA NUT: Called vegetable ivory, tagua nut is used today for carving.

TANZANITE (Indigo, Purple, Gold): Tanzanite is the gem form of thulite and zoisite (hydrous calcium aluminum silicate) listed in the epidote group. It additionally contains strontium and chromium, and is naturally colored violet-blue. It is found only in the Tanga province of Tanzania, Africa. Tanzanite embodies the highest mineral expression of peace on the planet, and through its power an individual is raised in vibration to the pinnacle of brotherly or agape love. Tanzanite should always be set in gold and preferably worn on the right hand or in the ears; if only one ear, the right.

Eons ago, after man crawled from the oceans and began his long development as a hunter/gatherer, some chunks of tanzanite were discovered in Tanzania which profoundly accelerated the mental transformation of humans. Man became aware of sea life, especially dolphins. A communion began on the shores of Tanzania in the present location of Dar es Salaam which enabled man to learn many advanced teachings beyond the scope his development would ordinarily allow. The chunks were transported around the world by dolphins in their mouths. The height of peaceful coexistence knew its flowering during the age we know as Lemuria, but as the teachings continued to flow, man's maturity was not advanced enough to keep forces in balance, and the age of Atlantis brought upheaval and an end to the early expression of the Cosmic Christ principle. Interactions with dolphins and tanzanite will produce immense positive emanations of peace power. Eventually tanzanite can be offered to dolphins without swallowing it, and he/she will transport the tanzanite where it must go or do with the tanzanite what must be done. Tanzanite is the bridge to Cosmic Christ consciousness which knew its first expression in Tanzania. Dolphins are an important piece of the linkage resonation to peace/love. Tanzanite is listed as an Auric Sealer in the Spiritual Apothecary. *(6th chakra)*

TARNOWITZITE: Tarnowitzite is a rare plumian aragonite. Some specimens come from Morocco. Tarnowitzite is a brain/nervous system relaxant. It allows creative persons to merge deeply with personal intent and vision. It is a soothing mineral. *(7th chakra)*

TEKTITE: Tektites forms as the result of a meteorite vitrifying or melting Earth's soil and extraterrestrial matter to create glass-like splashes. The clearest, highest example is Moldavite tektite.

Brown or Black - Tektite is comprised of various minerals affected by intense heat when it comes to Earth from other parts of the universe. It helps us link up with extra-terrestrials, and also tune into our own other world past lives. *(1st chakra)*

Green - See Moldavite Tektite.

Libyan Glass - Yellow stones are hard to come by mineralogically. The light and soft texture of yellow in this tektite provides numinous qualities, especially valuable for third chakra sensation of life purpose. Its off-planet meteorite origins further strengthen its utility in this capacity. *(3rd chakra)*

THOMSONITE: Thomsonite is a hydrated aluminosilicate of calcium and sodium classified in the zeolite group. It usually forms as globular radiating aggregates in small cavities of basalt. The white nodules have pink accents. Thomsonite is an excellent mineral for animals. It can be used in any situation where animals are kept captive to enhance feelings of serenity. This also applies to humans in prisons. It decreases anxiety and frustration about being closed in. It can be used by people suffering from claustrophobia to alleviate fear.

THOMSONITE and MESOLITE: This combination forms in white radiating crystals in half-spheres. Some are stiff, and some are fragile to the touch. These minerals in combination provide a guidance vibration which could be called the "Guide Stone." It activates latent sensitivities to reading environmental clues left by other living things, making it an excellent mineral for tracking. For people who act in a leadership or guide role, it provides strength, awareness and confidence. It is a good mineral to place in the sleeping area of guide animals, such as seeing-eye dogs. *(7th chakra)*

THULITE: Thulite is related to zoisite (hydrous calcium aluminum silicate), and is classified in the epidote group. Thulite can be a fairly pure massive, pink stone which contains high amounts of manganese, or it can be pink, green and white. Depending on the locality, the tones can be light to deep pastel shades. Thulite is a powerful heart chakra mineral. The pink/green combination exacts the highest level of heart chakra clarity in natural mineralization. This is true for any natural mineral which combines pink and green. Thulite's chosen path of union is as strong as placing an emerald and pink tourmaline at the heart center. Chosen union in this case indicates a high sacrament of mineralization, which is perennially meaningful to the spiritual activities of humans. Thulite is an exceptional stone to be given to newlyweds or for any anniversary celebration. It eases argumentation between any two people by opening the heart to clarity of emotional intent. It allows people to see the higher self of another. Thulite draws the rose of the lower part of the body into the green of the lung area, allowing the lungs to aerate the bloodstream and especially cleanse the lower body. Two people may perform a communion sanctity act by placing thulite in each hand and crossing the arms at the wrists, left hand palm up and right hand palm down, and clasp each other's hands with the thulite between each others' palms. This forms a cross-point, figure 8 (immortality), and sanctity spiral. This physically expresses two double helices in union, spiraling to Source which is powerfully uplifting. The essence of thulite is not to achieve healing through work or exertion, but to enact healing through the experience of sublime and sacred union. See Tanzanite and Zoisite. *(4th chakra)*

TIBETITE (Meteorite): Tibetite is a black, pock-marked splash of glass resulting from a meteorite strike in Tibet. Tibetite reduces blood pressure and other body functions. It is an excellent grounder, calmer and stabilizer. It fosters deep subconscious contact for therapeutic and psychic work. Tibetite deepens the connection to core body functions and individual frequency. *(1st chakra)*

TIGER EYE: Tiger eye is a quartz replacement of asbestos or brown iron oxide resulting from asbestos weathering.

Blue (Falcon's Eye or Hawk's Eye) - Hawk's eye is produced by inclusions of blue asbestos fibers. When one can see with the depth and clarity of a hawk, one is privy to much information about the world, the scene around them. It is an awesome responsibility to see with such incision (direct and decisive insight) without being overcome with empathy. This is the level of brotherly love putting itself on the warrior's path to truth, dignity, and honesty. If one is willing to see everything and still love and forgive, then one has made a magnificent step in the Christ consciousness. From the hawk's point of view, life without this deep vision is a life of poverty and dependence. The truest way for hawk to express divinity is through its unflinching, powerful gaze. The next most magnetic bird sight is the owl. The owl expresses ageless wisdom, a certain air of sage grace. The hawk is all fiery intuitive knowing, which activates the warrior spirit into action when needed. Only people with the hawk totem should wear hawk's eye, but at one time or another all are called to make the visionary transition into the Christ consciousness. This can be facilitated by meditation with the stone. Blue tiger eye is an Auric Opener in the Spiritual Apothecary. *(6th chakra and above)*

Cherry (Dark Red) - Cherry tiger eye is often heat-treated yellow tiger eye. Set the stage! Light the lights! This is an excellent stone for public oratory, presentations, etc. The qualities of this stone enhance confidence in oneself and the subject being presented. It allows clarity of thought to be projected with profound sensitivity to the needs of the audience. If the material being presented seems to be missing the point, it is not uncommon for the speaker to transition into descriptions which carry meaning and depth to the audience. With hematite inclusions, it is a sage combination, meaning that the wisdom of the total self is more easily brought into the dimensional reality of the user. This is a throat chakra mineral with resonance in the other chakras, thus individual harmonies are bridges to facilitate greater expression of self and the knowledge of the total self. *(5th chakra)*

Yellow (Yellow to Brown) - Its soothing blend of hues flow together like the words of one who wears it at the throat. The "mind's eye" can be focused, and insight is the natural by-product. It is actually quartz based. *(5th chakra)*

TIGER IRON: Tiger iron is a hematite-infused type of tiger eye, often with red jasper, from Australia. It is an extremely durable mineral combination, meaning that its vitality and usefulness will serve an individual without failure or complaint. Many minerals lose their efficacy because they stimulate growth and change which prepares the individual for a new octave of mineral stimulation. Tiger iron is one of a group of about twenty minerals which have the staying power to always be applicable. Tiger iron is an energizer and an excellent protection mineral. It promotes self assurance. It provides stamina. It is good for people who work at physically demanding jobs. *(1st chakra)*

TIN: Tin is a native element which historically was put to use as a coating for cans, because it is non-reactive and inhibits corrosion. It is a core-level healer, focusing on the lungs. This metal helps cleanse areas that have become inflamed and festered. *(5th to 3rd chakras)*

TITANITE: See Sphene.

TITANIUM ON QUARTZ (Auroraite™): "Laser stone." Titanium on quartz behaves like a laser for all physical and subtle body levels of energy work. The different colors expressed on the surface of the quartz, such as blue, yellow and magenta, allow this combination to address several chakras which can be determined by intent focused through the fingertips holding the quartz. It is naturally stable, and does not absorb negativity, disease or stress. *(All chakras)*

TOPAZ: Topaz is a hydrous aluminum silicate with fluorine. It bears a variety of colors; clear, yellow, blue, green, violet, and orange. It is a hard mineral, with a chisel tip, which has been substituted for in the gem trade by citrine. It bears a greasy luster and its crystals are stubby on top. Topaz is a great benefit metaphysically as a mineral which decreases visual fears, such as fear of the dark, scary images, and past memories. It aids eyesight physically, but is an essential mineral for any individual choosing to do difficult personal work or desiring to make progress in clairvoyance which is hampered by past trauma. Topaz is a Healer in the Spiritual Apothecary. Topaz is the state gem of Texas and Utah. Topaz is the November birthstone.

Blue - Blue topaz ranges in color from a light "Swiss" blue to a dark (heat-treated) "London" blue. Called "the Shield Stone" metaphysically, because blue topaz is long regarded as a stone of the protectorate whether police or military forces, but also of all activities which involve looking out for others or natural places, such as au pairs, nannies, forest rangers, guards and guardians. The wearing of blue topaz signaled one's occupation as a peacekeeper. It's cool blue tone aids in the attainment of clear and calm speaking, thinking and observation. Considered a stone which stimulates the high intellectual, analytical and rational mind, blue topaz is not given to psychic or intuitive stimulation. It is a stone which brings reinforcement to flagging will through the ability to problem solve and conceptualize. It strengthens intention and attention to detail. *(5th and 6th chakras)*

Brown - This form of topaz is applicable to colon cancer, colitis, Crohn's disease, and other ailments afflicting the first and second chakras, especially dealing with the elimination system. It strengthens second and first chakra organs and tissues. It is relaxing to the muscles of that region aiding to counteract constipation and cramps. *(1st and 2nd chakras)*

Champagne - This quality of topaz is highly uplifting. It induces optimism, euphoria and ecstatic visions. It is specifically beneficial for lymphatic diseases and brings a relaxation of dread or fear regarding difficult diseases such as leukemia, lymphoma, and bone marrow cancer. *(4th chakra)*

Imperial (Orange to Golden) - This form of topaz works with the digestive system to aid assimilation of nutrients and expression of energy. The direct will center impact facilitates correct action. It promotes well-being and a consciousness that prosperity is having a sufficient amount for comfort rather than a gross abundance which exceeds comfort and becomes obscene consumption. It is useful to assuage hunger. *(2nd and 3rd chakras)*

Red - A very deep brown/orange, this quality of topaz addresses ailments and diseases of the reproductive organs. It also instills earth fire into the first chakra, and is an excellent counterpart to support bringing in earth energies during healing work. *(1st chakra)*

White - This gem quality topaz opens the crown chakra to spiritual fire. It tends to cause deep relaxation as a response to the opening, therefore it is best worn as a gemstone with stones that enhance grounding. Some white topaz can be found which has black tourmaline included in it, an indication that earth knows what, when and why to combine. White topaz, because of its fluorine component, is useful as a psychic protection mineral. *(7th chakra)*

TOPAZOLITE: Topazolite is found in small crystal clusters, and provides deep cleansing support, both physically and non-physically. It can be used to dissipate mental fixations or obsessions, though it should be complemented with a calming mineral to complete the release process. In physical work, expect an emotional reaction and be prepared to use calming minerals. *(1st chakra)*

TORBERNITE: Torbernite is a hydrated copper uranium phosphate classified in the autunite group. It crystallizes in flat, bladed crystals of light green color. Torbernite is an excellent mineral for people who work continuous stretches of 24 or 36 hours. It has a mobile ability to draw energy reserves to the body from surrounding sources, such as light or water. It provides strength to maintain focus under stress. *(All chakras)*

TOUCHSTONE (Lydia Stone): A siliceous black rock, sometimes altered to gray, also called lydite in England. In ancient times, used for streak (scratch) testing gold or other precious metals. The trained eye could judge the degree of purity by the coloration left. Basinite is a current equivalent. *(1st chakra)*

TOURMALINE: Tourmaline is a hydrous borosilicate which contains other elements depending on its type. The specific types listed below indicate the mineral name associated with the tourmaline group classification and its color. In its color range, tourmaline activates the associated chakras from black (schorl), first; red (rubellite), first and second; brown/orange (dravite), second and third; green (elbaite), pink and watermelon, fourth; blue (indicolite), fifth; cat's eye, sixth; clear, seventh. Generally, it is an emotional stabilizer, and it draws off static and tunes the nervous system to endure large amounts of incoming light. Tourmaline as a family holds a masculine frequency. Tourmaline is a ring silicate of molecules stacked one on another in sheets held together by boron atoms. Tourmaline was used in submarines for high pressure gauges, because the boron in its molecular structure provides enormous strength on the physical level. Pyrex is a borosilicate glass which can withstand rapid changes in temperature without cracking. Tourmaline expresses strong characteristics of piezoelectricity and pyroelectricity.

Metaphysically, tourmaline and danburite are brothers sharing the distinctly useful attribute of supporting huge amounts of light energy to be withstood by the subtle and physical bodies. Tourmaline is the state mineral of Maine. Tourmaline is the October birthstone. All forms of tourmaline are Healers in the Spiritual Apothecary. Of all the choices available in the Healers division, all forms of tourmaline are desirable second only to morganite. *(All seven body chakras represented)*

Dravite (Brown) - Dravite contains sodium, magnesium and aluminum in the borosilicate matrix. Dravite forms in stubby, one to two inch crystals with flat sides making it look like an elongated garnet. Strongly associated with healing and cleansing in the will (orange) chakra. Excellent for the legs as well. As the person lies on the back, place dravite, on the stomach area. For hip problems, roll on either side and place on hips. Minerals which work well in unison with dravite are "halo rock" (selenite) and jade. Selenite draws white healing light through the crown chakra and through the chakra chord. Jade can be placed on the heart chakra to ease tonal frequency updrafts and downdrafts, while the selenite and dravite interact. This stone is particularly useful to people who gather their illnesses in the will chakra. Dravite earns its own recognition as an Etheric Cleanser in the Spiritual Apothecary with attributes it specifically offers as unique from schorl (black tourmaline) which is also in this category among others. *(2nd chakra)*

Elbaite (Bi-Color, Watermelon, Green and Dark Pink to Red) - Elbaite contains sodium, lithium, and aluminum in the borosilicate matrix. Pink and green tourmaline is best described by the term "solipsism," which is a theory that says the self can only know its own changes or modifications and that the self is the only existent thing. The grace of pink and green together in the mineral realm is that they both relate to the heart chakra. No other chakra has two such different colors relating to it. Bi-color tourmaline is the highest expression of unified self, and this relates to all beings and all awareness, not just individuality. The beauty of bi-color tourmaline is that the only existent self is recognized as the entirety of the cosmos. Using it supports acting in integrity with this higher awareness. *(4th chakra)*

Elbaite (Green) - Elbaite contains sodium, lithium, and aluminum. It increases courage and the strength to cope. It was used in the past for couples separated by war or the hunt. It provided benefits to both the male and female. There are both green and pink versions of elbaite. *(4th chakra)*

Indicolite (Blue) - Indicolite, normally the most rare of tourmalines, is probably the purest form of tourmaline. Blue tourmaline is unsurpassed in supporting connection with higher self and advanced consciousness. Earth's sense of self began as a water consciousness and her highest beings are water mammals, especially cetaceans. As she grows continents, she becomes increasingly aware of land-based consciousness. Using indicolite fosters a strong connection to our water heritage and eases linkages with Gaia consciousness. *(6th and upper chakras)*

Rubellite, Variety of Elbaite (Pink) - Pink tourmaline supports the attitude of being young at heart as well as selfless service to others. As described under bi-color tourmaline, pink tourmaline places the individual in a state of readiness to feel connection with the entire cosmos, but pink is oriented always outward. There is no sense of individuality and being all. Therefore, pink tourmaline can be overused in modern society by individuals who are not aware of the service implications of this mineral. It places a strain on them to move outward which can be confusing but beneficial in the long run. *(4th chakra)*

Rubellite, Variety of Elbaite (Red) - This red tourmaline, often found combined with lepidolite, produces a softly passionate influence. It activates love and devotion in a balanced way. *(1st and chakras)*

Schorl (Black) - Schorl contains sodium, iron and aluminum in the borosilicate matrix. It forms much like dravite in stubby, flat sided crystals. Schorl is excellent for grounding, inducing deep, restful, dreamless sleep, and for cleansing the physical, auric and etheric bodies. It draws like salt or halite and fluorite. It is protective like obsidian. Unto itself, it is a harmonizer, promoting deep and clear serenity and safety allowing individuals to come together equally to achieve desired goals or results. Schorl is a unique mineral in the Spiritual Apothecary, because it occupies five categories: Auric and Etheric Cleansers, Auric and Etheric Sealers, and Healers. Its capacities are broad, and it possesses the matrix to promote spiritual surrender through uncommonly deep meditation support as well as the opposite of conscious, mundane and business-like focus during daily activities or especially times which require intense presence of mind and focus, such as during job interviews, contract negotiations and signings, and public presentations. Schorl draws white light energy deep into the base chakra, because the boron in its matrix allows strong light currents to enter the body without causing physical distress. Schorl is the Hercules or Artemis of minerals, but its humble quality makes it nearly invisible as a light worker. It is the most important tourmaline for overall healing purposes. Since it is part of a family which is naturally masculine, it supports moving forward spiritually and causally. Schorl is the antithesis of couch potato. *(1st chakra)*

Uvite (Black) - Uvite forms in small, flat, bladed crystals. Black uvite is a tool for spiritual practitioners, especially people who make flower essences and tinctures, people interested in space travel and extraterrestrials, and people who provide past life information to others. It is important to work with this mineral consistently over a period of time in order to understand the variation of information it facilitates receiving. *(1st chakra)*

Uvite (Green) - Green uvite forms in small, flat, bladed crystals. Uvite stimulates peristalsis, the involuntary contraction of muscular structure in the intestines. It also rehabilitates degenerated glands all over the body. Uvite is an excellent antidepressant. *(2nd and 3rd chakras)*

Uvite (Red) - Red uvite forms in small, flat, bladed crystals, and is a foundation stone revered in the past as capable of aiding geomancers in orienting buildings and selecting appropriate building sites. It also supports setting guidelines for new projects which are implemented by several or many people. *(1st chakra)*

TREMOLITE: Tremolite is a fibrous form of jade classified in the large and complex amphibole group of minerals. When pure, it is white, but impurities will darken it slightly green. It forms a solid solution series with actinolite which is the parent mineral of nephrite jade (not to be confused with pyroxene jadeite). Tremolite is a hydrous calcium magnesium silicate. Tremolite is a subtle body stone. It moves the user from physical consciousness into consciousness of the next dimension. All forms of tremolite help to stimulate creativity and imagination after periods of grief and loss.

Green - Gives practicality to creative plans and ideas. It focuses attention on what is possible. It works primarily at the physical level. It aids in reestablishing neuro-ending connection or restoring frayed nerve endings. It helps to focus energy on what is possible. It works well with any green stone. Green tremolite, especially, like other amphibole group minerals, resonates to frog medicine. *(4th chakra)*

Pink/Lavender - Aids in the improvement of love relationships. It activates imagination and creativity. It soothes the nervous system with lavender or pink light. It works well with sugilite, particularly when holding one stone in each hand during meditation. *(4th and 6th chakras)*

White - White tremolite repels and dissipates negative thought forms. It integrates the higher mind with the chakras above the body, and improves learning mental processes such as biofeedback. White tremolite is the most neutral form of jade, therefore it is excellent to support detaching from grief and loss and seeing the higher order of choice in determining karma. Tremolite bathes the nervous system in white light. *(7th chakra)*

TRINITITE: Trinitite, atomic bomb glass, was formed in the first atomic bomb explosion on July 16, 1945, near Alamagordo, New Mexico at the Trinity Site. It bears a great sadness within it because of its origins; however, it bears within it a fundamental shift in consciousness regarding the Trinity of Father, Son and Holy Spirit. Historically, the essence of the Holy Spirit has been ill-defined by Christian scholars and teachers, but it held a definite masculine tone. Through this tone, the Holy Spirit regains her rightful identity as the Earth, the Great Mother, Mother of All. Through that atomic moment in particular, her consciousness changed regarding humankind. *(2nd, 5th and 8th chakras)*

TUNNELITE: "Stonecypher stone." This hydrated strontium borosilicate surpasses selenite's ability to open the physical vehicle to electromagnetic vibration allowing clear receivership and transmission on the psychic level. For people seriously involved in mineral healing and translation of mineral information, this mineral enhances intellectual understanding and intuitive connection. *(5th through 9th chakras)*

TURGITE: Also called hydrohematite, turgite is a hydrated iron oxide which probably is a mixture of hematite and goethite. It is good to dispel anxiety attacks, and brings rainbows back to those who have not been in the world for a time, such as people experiencing agoraphobia. This mineral is an excellent stimulator for depressed people. It is an excellent stone to enliven homes which have experienced death or loss of a loved one in divorce, etc. It is a home-oriented stone, being more conducive to privacy than to business world or social matters. It provides an acceptable balance between needing safety and privacy and moving out of that space with a sense of confidence and assurance. *(1st and Auric chakras)*

TURQUOISE: Turquoise is a hydrated copper aluminum phosphate which can be quite useful in balancing electrical issues in the body. It combines well with iolite and calcite to address many physical issues. Turquoise is the state gem of Arizona, Nevada and New Mexico. Turquoise is the December birthstone. It is located in the Healers category of the Spiritual Apothecary. See Richterite.

Blue - Longevity is increased by the ability of turquoise to protect the primary body systems from pollution. Therefore, it is an excellent stone for children as they are developing their immune systems and older people. Overall toning physically and emotionally gives its wearer vibrancy. It is considered by several cultures to bring luck and prosperity. It counteracts depression and balances mood swings. It provides the correct elements for energy to be utilized fully for mental clarity and concentration as well as increased oxygen carrying capacity in the blood afforded by copper's effect on iron. Selected as a sacred stone by the Native Americans, its list of attributes could fill a book. It represents the masculine principle or father sky, and many homes in the southwest have doors painted blue as a complement to the sky on earth. Turquoise is readily available in the southwest United States as several of the world's classic locations are found there. It is also found in Tibet. These planetary locations indicate that turquoise is effective in transmitting a ray of peace and harmony which influenced the Hopi and Tibetans. It induces high levels of mental efficiency which encourages greater connection to divine principles. The fact that both cultures live at higher than average altitude increased their likelihood of benefitting from celestial incoming frequencies. They utilized the mental clarity and oxygen carrying capacity effects of turquoise to their optimum for spiritual development. Turquoise is found in the Spiritual Apothecary in the Healers division. *(5th and 6th chakras)*

Green - Like blue turquoise, green turquoise is an excellent healing mineral. The green ray tends to be more activating for physical needs, and surpasses the blue in providing luck. Green turquoise is a wise person's stone, a stone for old souls. It supports reclaiming past life prowess and training. *(4th chakra)*

TURRITELLA: A fossil composite. See Agate.

TURTLE BACK (Stone): "Logos or The Word stone." This stone originates from the rainforests of the Olympic Mountains in Washington state, where emanations of the Great Mother are kept strong by the tremendous powers of earth and water in this area. The Olympic Rainforest is the only temperate rainforest in the world, because storms from the Pacific, the largest ocean in the world, drop an average of 180 inches of rain a year. Originating in a region of such intense rainfall, turtle back literally represents the cleansing of the earth, soul, spirit, physical body, psyche, heart and mind. It is a good stone to use in other regions to connect with rain spirits. Turtle back, representing the foundation of life on earth, is the body of the Mother, the womb of life. This stone, when given to expectant women or a mother, supports her role as embodiment of the Divine Mother. It is suffused with the Christian concept of Creative Word or Logos, so placing this stone on altars and outdoor sacred areas syncretizes or brings together diverse energies, beliefs and elements. It composts and reinvigorates traditional ways to be used a new in the light of present needs and new awarenesses.

Turtle back is a fuser as in the sacred moment of fertilization. It embodies goddess energies. Turtle back stone offers a sense of grounding in the storm, of protection during the time of earth changes. It encourages waiting for the time of power, authority and speaking one's story as an authentic result of experience and being recognized by the community for maintaining a steady, methodical, trustworthy course. Turtle back stone is stabilizing, and encourages going within to attain spiritual depth, indestructibility, harmony and silence. Turtle back embodies cosmic harmony and purity through its associations to the lotus flower. Turtle back is a great complement stone to ocean picture stone as turtle is the support for elephant. See Nunderite and Ocean Picture Stone. *(1st and 4th chakras)*

ULEXITE (TV Rock): Ulexite is a sodium calcium borate very rarely forming in crystals. It looks like an aggregate of fine white, hair-like fibers. It is extremely soft (hardness of 1) and fragile. Like kernite, ulexite is an ore of boron. Crystals of ulexite can be found in Boron, California. When it is cut cross-grain and polished, this mineral is transparent and produces a "3-D" affect when looking through it at print. It is an energizer, and is good for warming up the nervous system connection to the muscles prior to warming up the muscles for physical exertion. This nervous system connection can help to alleviate nervous disorders such as tics and nervous system diseases such as muscular dystrophy. It helps to balance out the overactivity of hyperactive people, and induces calm. It is an excellent grounder for hyperactive people. Ulexite is a neutral energy shamanic power mineral. *(All chakras)*

UNAKITE: Of the family of pink/green minerals, unakite is the elbow of the body. The elbow is a necessary junction in the arm, allowing movement and flexibility to the limb; but it also gets bumped because it is willing to be located at a focal point of activity. This analogy shows the inherent use of unakite as a heart chakra mobilizer. Often, when other minerals and mineral combinations have helped to heal the heart chakra, there seems to be no movement or new growth, or the willingness to be bumped, to be out there in the activity of the world. Ukanite will support movement, and is the obvious follow-up to a series of heart chakra healing applications. This phase of green is highly grounded, and provides the heart knowledge that a firm base of intuitive power has been revealed through healing and release. This supplies the heart chakra with confidence. This phase of pink is strengthening and motivating. It provides vision to the newly exposed heart. When all seems bright with expectation and promise, unakite says, "Yes, by the grace of God, follow your heart. Be your love. As you are, stronger shall you be." Unakite embodies the masculine aspect of beaver energy. Unakite is an Etheric Sealer in the Spiritual Apothecary. See Agate (Burro Creek). *(4th chakra)*

URALITE: Uralite is comprised of rectangular, stocky, and dull green crystal nests. It is primarily found in Salida County, Colorado. An interesting mineral for teachers for it promotes awareness of the learning needs and natural talents of others. It enhances the development of self esteem in healthy ways. Not usually useful alone, however some highly evolved beings benefit from its presence because it dispels negativity at a very weak frequency which usually is not powerful enough for most individuals.

URANINITE: Uraninite is uranium oxide. It is sometimes pitch black, but can be associated with other minerals. The light violet to dark purple range of color stimulates the third eye and crown chakras to promote intuitive and spiritual opening. This mineral deepens the sense of Thy Will Be Done. *(6th and 7th chakras)*

VANADINITE: Vanadinite is a vanadate and chloride of lead classified in the apatite group. It is usually colored a deep brick or blood red in brilliant stubby crystals. The best specimens are from Morocco. The course texture of this mineral's name suggests a strong, powerful presence. As an apatite group mineral, vanadinite is the highest choice for providing backbone to a decision which has been difficult to reach and is anticipated to be difficult to implement. This mineral needs to be honored and activated with extreme reverence, caution and discipline. It markedly enhances the intention of the wearer/user. It brings invincible power to correct action. One is cautioned to carefully pray about and meditate on intent prior to activating this mineral. It is not a mineral to be placed in the general environment as it will work to manifest all intentions released in its vicinity. Only a skilled practitioner would take the responsibility of having this stone highlight every nuance of every decision. Vanadinite is a powerful expression of the archetype of the divine father. Vanadinite embodies armadillo energy. See cavansite and descloizite. *(1st toward 3rd chakras)*

VARISCITE: Variscite is a hydrated aluminum phosphate, which forms as a light green and white nodule. It supplies energy for healing to the DNA level. It passes unwanted energy into the blessed Earth when placed at the first chakra during a healing application. Variscite can be worn continuously anywhere on the body. Variscite supports many functions in the body from protein to lipid production as well as energy release. It is found in the Spiritual Apothecary in the Healer division, and is an important mineral for shamanic and energetic healing practices. Variscite combines well with iolite. See Richterite. *(1st to 4th chakras)*

VERDITE: Verdite is a green, mottled carving stone from Africa which contains serpentine, corundum and other minerals. It works to literally verify "hunches." It is an excellent mineral to use to back up intuitive powers. The stone, upon asking a question, will provide an electrical impulse which activates the "truth response" in people. Most people will experience a tremor or a chill sensation going from head to toe. Depending on the level of verification provided, the impulses will range from one to nine. No sensation at all indicates no verification. Some people will experience paroxysms or shudders, and no specific number of impulses are provided. The person will convulse either mildly or strongly, depending on the verification response. Through use the person will learn to gauge the level of verification from the experienced electrical impulse. *(4th through 6th chakras)*

VERMICULITE: Vermiculite is a group name for silicates containing magnesium, iron and aluminum. It is a soft, yet crispy mineral which forms in bands of tan, cream and gray colors. It is used industrially as an absorber. For the physical body and other minerals, it is a good absorber of toxins and pollutants, and needs to be kept clean prior to use. Vermiculite can be used like soapstone to cleanse other minerals. It tends to lighten the physical body, so it is a good mineral for people with excessive amounts of earth signs in their horoscope.

VESUVIANITE: Vesuvianite is a hydrous calcium magnesium aluminum double silicate. As if one silicate molecule was not enough, this anomaly allows vesuvianite to offer the highest embodiment of Goddess energies found in all mineral forms. All forms of vesuvianite are located in the Healing section of the Spiritual Apothecary under the name Idocrase, which is an old-time name for vesuvianite.

Black Idocrase - Idocrase is the necessary complement to danburite when performing psychic surgery. It grounds the healing practitioner to earth energies which flow through the practitioner, to star energies, which are connected to the person being worked on. For individual use, idocrase is excellent at the feet for a meditation layout done lying down. Goshenite at the head and dioptase at the heart with idocrase at the feet is an excellent combination of minerals for deep cleansing and relaxing meditation. This form of vesuvianite resonates to the Black Madonna or Dark Goddess. *(1st chakra)*

Brown/Orange - Vesuvianite is a powerful second chakra healer. It was referred to as the surgeon's apprentice in ancient China because it was commonly used as a support stone for obsidian in surgery. It is an uncommon totem mineral, excellent for those working on will expression from the second chakra. This form of vesuvianite resonates to the Hawaiian Goddess Pele and to all goddess energy worldwide which creates from her matrix. *(2nd chakra)*

Green (Californite) - Green vesuvianite is found from California to Washington on ocean beaches. It forms in spotted green and black of different hardness which causes pea-sized dents all over chunks of it as the colors change. It is a whole body healing mineral, and green vesuvianite offers excellent properties for calming the entire body system for meditation, sleep or hypnosis. The ancients, our ancestors in spirit, call green vesuvianite the "Star of Shakti Astri," a distant galaxy named by its central star.

Legend said the galaxy was the home of the goddess Isis who came to Earth intending to find her sister Astarte, the Phoenician goddess of fertility and sexual love. Her curative and fecund powers were transferred to green vesuvianite when she returned to Shakti Astri with Isis. Astarte was worshipped in many cultures as the queen of the stars, and by many different names, all generally meaning star. *(All chakras)*

Lavender and Green (Crystallized from Jeffrey Quarry, St. Hilaire, Quebec, Canada) - A mineral of divine qualities. It is called by our ancestors the protectress of Shakti Isis, or the seat of Shakti Isis, for Isis is the creator, the mother of God, and from her all life flows. It was said when she departed to Shakti Astri with Astarte that her essence transferred to lavender vesuvianite. Using both green and lavender vesuvianite in meditation or healing work is especially powerful. The resonance is unequaled. *(6th and 7th chakras)*

VICTORIASTONE: This is a manmade amalgam of minerals, including actinolite and other jade-like components. It has the texture of glass. Magenta, lime green and deep blue are the most common colors. It is an excellent stone for color therapy applications, and good for gazing to enhance visual impact. The chatoyancy is pleasing and mesmerizing. Chakra effect varies with color.

VIOLANE: This is diopside with manganese, and is purple, making it an excellent spiritual companion for prayer, contemplation, or meditation. Violane can be greatly beneficial in bringing periods of subsidence to the first stages of grief where a person can rest either in a warm bath or in a softly sedated state. *(6th chakra)*

VITRIOL (Blue): See Chalcanthite.

VIVIANITE: Vivianite is a hydrated iron phosphate classified in the vivianite group. It is rare in crystalline form and usually colored green or deep blue. It is normally transparent when first mined, and darkens when exposed to light. It is fairly common as a pale blue powder distributed in peat bogs, but when it crystallizes, it forms in blocky, flat-edged crystals. Vivianite is peerless. It is the entrusted stone used in many capacities by trained and knowledgeable women. It is balanced by feminine energies so it is best used by women, though its energy is not exclusive to women. The spire shape is something of an anomaly, and should be wrapped and carried carefully.

One important aspect of this mineral is that it is a "space writer." The thought form is placed in the air by the handler, and can be left as a message (to be viewed when a match is burned), or can be transmitted across space/time - something like a plane trailing letters or sign language. As with all minerals in the vivianite group, vivianite enhances a form of communication. In this case, the communication is silent and transmitted through the hands, so this is an excellent companion stone for the deaf and mute.

It is a perception healing tool, and to the sensitive, acts as a stethoscope, allowing the other hand to become more sensitive to tumors, breaks, etc. The cluster form of vivianite is excellent for placing under a diagnostic table, and was often employed in this manner, aiding the healer to read what ailed the patient. In a room setting, it helps to dispel emotional charge which is disruptive or counterproductive, and is excellent for enhancing communication on all levels. It is a stone of blood ritual such as first menstruation, childbirth, moontime ceremony, the Sun Dance, and communion. Vivianite is excellent as an aid to evaluating depth of involvement and commitment. For people who feel obligated to work for others, such as mothers, vivianite strengthens self-care and holding others able to do their share and take care of themselves. It is located in the Spiritual Apothecary in the Healer division. See Richterite. *(6th chakra)*

WAD: Wad is a common name for a group of hydrated oxides of ores such as manganese, copper, and iron forming in a compact flow of brown to black masses with some luster. Asbolane is a form of wad which contains a high amount of cobalt oxide. The oxidizing conditions which create wad occur in bogs, lakes or other shallow water environments. Manganese samples of wad stabilize fragile nervous systems, and reduce anxiety, impatience and intolerance. It invokes a gentle awareness of others' needs and tends to balance male hormones, particularly testosterone. It is useful to people who are easily distracted and overstimulated (ADHD). It is grounding. Wad neutralizes overly acid conditions, and helps health conditions resulting from excess acidity. *(1st and 2nd chakras)*

WARDITE: Wardite is a hydrous sodium aluminum phosphate. Wardite forms in pea-sized, creamy white crystals. The best specimens are found near Rapid Creek in the Yukon. Wardite is a mineral in tune with the jet stream. It can unleash the awareness of one's own cruising speed in the rest of the sky allowing an individual to find that slip stream of self leading to spiritual unity. *(7th chakra)*

WAVELLITE: Wavellite is a hydrous aluminum phosphate fluoride which forms in light green botryoidal bumps on a gray rock. A well-known deposit is found in Arkansas. Wavellite brings in a strong quality of discernment skill. Any sort of work where the user must enumerate qualities or aspects of a thing is enhanced. It is a Virgo-type mineral, and melds with people who have Virgo in their horoscope; however, its greatest benefit is derived by those without Virgo analytic or discernment skills, as it provides this support to individuals. Wavellite emphasizes the ability to view others dispassionately. It dispenses with anger, and allows clear and cool appraisal. It strengthens the ability to see the truth in others, and the willingness to celebrate them as they are without need of changing them. This is a mineral which praises diversity and reduces animosity and prejudice. *(3rd and 6th chakras)*

WHITEITE: Whiteite is a name for three minerals (all called whiteite) classified in the whiteite group which are magnesium aluminum phosphates. The types of whiteite contain either: 1) calcium, iron and magnesium; 2) calcium, manganese and magnesium; or 3) manganese, iron, and magnesium. It forms in blocky, off-white crystals found in the Yukon. The two types with iron in them are extremely hardy minerals which promote survival skills, tracking, and keen awareness in the wild. They strengthen feelings of safety and respect in the wilderness, so individuals do not become destructive out of fear of or greed for natural surroundings and resources. The type with calcium, manganese and magnesium helps to maintain special orientation in strange places or the wild. *(1st and 2nd chakras)*

WILLEMITE: Willemite is a simple zinc silicate. It is found massive with franklinite in Franklin, New Jersey, and fluoresces green and orange. Known as "Signature rock" metaphysically, willemite facilitates spirit orientation to genetic coding. It is called signature rock, because at the spirit level, each physical being has a characteristic molecular spin, which defines exact and distinct polarity and magnetism. Using willemite greatly increases the success of psychic healing on all levels. It can be a fundamental component of all spiritual apothecaries, and is listed in the Healer division. *(All chakras)*

WITHERITE: Witherite is a barium carbonate classified in the aragonite group. It forms in modest, creamy tooth-like crystals. Its yellow/white color gently stimulates the third and above chakras encouraging acting instinctively, from one's center of awareness. It stimulates immediate translation of thoughts into action in the way that thoughts must become deeds. It supports sacral and solar plexus connection to the higher chakras. *(3rd through 7th chakras)*

WOLFRAMITE: Wolframite is an intermediate member of the heubnerite-ferberite series. Heubnerite is a manganese tungstate and ferberite is an iron tungstate. Wolframite forms in heavy, thick tabular crystals with striations. The best specimens may originate in Panasquera, Portugal. Wolframite is most effective in combination with chiastolite or andalusite. Without either, wolframite tends to be disorienting to human circuitry. With either of these other two minerals, wolframite synthesizes and synchronizes discordant brain wave frequencies, allowing clearer deductive thinking. Excellent for higher left brain functions; however, this combination also strengthens right brain abilities. *(6th chakra)*

WOLLASTONITE: Wollastonite is a calcium silicate, with at least six polytypes known of this mineral. When found in massive rock, wollastonite forms in white, brush-like crystal patches, mixing with garnet and diopside. It integrates the first and seventh chakras, and is excellent for meditation, balancing and chakra cleansing. It enhances creativity. Wollastonite is a good mineral combination for spiritual ceremonies. *(1st and 7th chakras)*

WULFENITE: Wulfenite is a lead molybdate. If often forms in square or rectangular, clear tabular crystals under an inch in size. This is a soft mineral much prized by collectors. Wulfenite is an Etheric Conditioner in the Spiritual Apothecary.

Butterscotch - This vibration of wulfenite stimulates absorption of vitamins and minerals, and is useful for the aged in toning the metabolism of all nutrients. *(3rd chakra)*

Red/orange - Yellow, red-orange clusters of flat crystals act on the navel area, activating the spleen and organs of digestion. Placing or carrying this mineral in this area helps clean and purify the system as toxins are dislodged during the energizing of the abdomen area. *(2nd chakra)*

YOUNGITE: Youngite is a chalcedony with drusy quartz from Wyoming, now extinct in its original location. It fluoresces green. This furry-skinned (drusy quartz) mineral looks like enfolded coral with a coat of small crystals covering a reddish-brown matrix. Its fluorescence allows it to retain light to illuminate darkness. This can be enhanced by color infusion treatments — blue light is well suited to this mineral. Holding this mineral in the dark with ultraviolet light and staring at it allows its inherent sparkles and streamers of light to come to life (much like the "night sky" visual obtained in a closed-eye meditation). Working with this mineral opens the third eye and facilitates night vision, but also stimulates the throat and heart chakras. Youngite is a mineral of wisdom, contemplation and devotion, all stemming from its resonance with the heart. It is a mineral of hidden virtue, of holding one's counsel until asked, and watching without judgment. *(4th, 5th and 6th chakras)*

YOWAH NUT (Fossilized/Opalized): Yowah nut is a higher form of boulder opal. This mineral is an Earth speaker. It is imbued with an understanding of Earth cycles, life kingdoms and processes. It is invaluable as a medicine stone. When taken on journeys, it provides information about localities which could be used immediately for finding food and shelter.

ZEBRA STONE: Zebra stone is between a sandstone and jasper. It is found in Australia and is striped in elegant tones of tan and brick. This two-tone painting by the Great Mother instantly projects and makes visible the aboriginal energy of its homeland. A gentle centering and grounding provided by its earth tones is not a weight. When you want to perform higher spiritual work but stay earth-connected, this is the stone to carry. *(1st and 4th chakras)*

ZEKTZERITE: Zektzerite is a sodium lithium zirconium silicate. Zektzerite has shamanic powers. It is capable of dispelling static energies around a body as smudge does. It is excellent to pass over deceased bodies prior to burial as it negates unwanted energies so the spirit can see clearly to leave the body and proceed to the next level. *(Auric chakras)*

ZINCITE: Zincite is a zinc manganese oxide found crystallized in old zinc smelter chimneys in Poland. In its clear, lustrous state, this mineral's coloration spans from blue-gray to light green to deep yellow to orange-yellow to dark red. It is heavy. It is a powerful circulation system stimulant, and bears the role of "engine" in many situations quite well. It can be applied to frostbitten areas without negative effects of too rapid heating, and it is excellent to apply to the second chakra for hypothermia and shock due to drowning or cold exposure. Zincite works throughout the body to strengthen the immune system, and has a positive effect on all lymph nodes. It increases resistance to pathogens. It helps to counteract laziness, sluggishness, chronic fatigue, sterility and impotence. It aids digestion and the liver as well as vitamin utilization (especially the B vitamins).

Zincite announces clearly that it holds space. It supports self-esteem, personal boundaries, and recognition of toxic patterns, situations and people. It supplies energy to implement changes in one's life, and distinguishes itself as a powerful ally of the human spirit. It transmits the energy to separate from tribal influences to make the leap of faith into the unknown territory of one's own path. That zincite originates in Poland, the first Soviet block country to fight for independence, says a great deal about its properties and importance to social justice. *(2nd and 3rd chakras)*

Blue/Gray - This shade supports the thyroid and lymph glands in the neck to counteract strep throat, laryngitis and upper respiratory ailments. It supplies great energy to speak eloquently for the rights of others and to do so with charisma and dignity. This shade stimulates the dawning awareness that spiritual truth, in order to be engaged, affects the whole body. This chakra must be a safe conduit for Divine will to enter the lower chakras and for personal awareness to rise to enter into full presence of the Divine. Zincite provides this chakra cleansing and vibrations of willingness and safety. *(5th chakra)*

Green - Green zincite is especially useful in aiding respiration of cells which involves the intake of oxygen and the expulsion of carbon dioxide and waste products. This shade especially aids the liver, heart and thymus gland. This shade calms the nervous system. It helps the heart to disengage from troubling or painful situations, and it moves out toxic emotions to make way for compassion, forgiveness and love. It helps the energy and power of love to be fully engaged. It reduces crippling self-consciousness and shyness, and deepens the relationship to self. *(4th chakra)*

Orange - Orange zincite aids in production of male hormones, and it provides enormous energy to the body. It can be thought of as the salvage operator in the body, gathering up all unnecessary elements and moving them out through the elimination system. This shade provides a sense of control and mastery over choices. It helps to reduce things to their essential usefulness, otherwise they are let go. It increases the sense of safety in intimacy, and provides energy to engage deeply with others. *(2nd chakra)*

Red - Red zincite provides the properties of yellow and orange zincite as well as base chakra healing power to address venereal, bladder, and testicular diseases. It specifically charges the lymph nodes in the groin with healing energy. Red zincite deepens the ability to question whether loyalty is appropriately placed, and if not, the stimulation to adjust one's convictions. It enhances the sense of being in the right place, liberating the psyche and emotional body from stress and anxiety. *(1st chakra)*

Yellow - Yellow zincite aids the pancreas in insulin production and in the storage of glycogen, which is the primary carbohydrate storage unit in the body. Yellow zincite begins the process of carbohydrate breakdown which is completed by orange zincite. This shade calms the nervous system, and it is the shade of self-esteem. It allows an individual to separate from criticism and rejection. It supplies a sense of safety to psychic awareness originating in this chakra as well as invigorates zest and inspiration for life. *(3rd chakra)*

ZIRCON: Zircon is a zirconium silicate. In its clearest, most gemlike quality, zircon stimulates spiritual knowing, and provides emotional protection. An individual can become a truly independent force by tapping this mineral's energy. Zirconium is quite durable, and aids an individual who feels easily overwhelmed or overpowered to develop boundaries and thresholds of openness to others. It is a useful mineral for empaths and psychic sensitives as it helps restore depleted electrolytes. Zircon is the December birthstone. *(All chakras)*

ZOISITE: Zoisite is a hydrous calcium aluminum silicate classified in the epidote group. It is dimorphous with clinozoisite. Zoisite's striated crystals are found in brown, blue, green, and white. When found with ruby, it is green flecked with black, and called ruby zoisite. This form of zoisite has a profound impact on the root chakra. With better vibratory flow in this area, a person's overall energy level and adrenal balance are enhanced. Both the male and female sexual organs are supported to health, and increase in fertility is natural. It connects deeply to the etheric subtle body as well. Zoisite, like others in its family, promotes intimacy and safe contact in touch. It is an excellent stone for reducing touch sensitivity in autism and sexual abuse recovery. It soothes the skin layer of the body, increasing sensations of well-being, trust and safety. It greatly increases awareness of skin as a component of visual intake. Zoisite affirms convictions rising from body sensations, and enhances the ability to form images or concepts cognitively to make sense of incoming vibrational information. At its peak of clairvoyant support, zoisite brings awareness deep into the body, so one is able to decipher clearly, and say, "I just knew it in my bones." Zoisite is a stone of deepest reckoning brought to consciousness through the heart and root chakras. Zoisite bears a feminine vibration. A pink variety with manganese is called Thulite. A violet variety with chromium and spontium is called Tanzanite. See these minerals as well as Epidote. Zoisite is listed as an Etheric Conditioner in the Spiritual Apothecary. *(1st, 4th and Etheric chakras)*

APPENDIX 1

Primary Chakra Aspects

On the following pages, minerals are arranged by body zone indicating which area or chakra their energies impact based on the chakra influences noted at the end of mineral readings in the previous section. Minerals open and close these energy centers through resonant vibration, and optimize release and flow of healing energy inward.

The term "chakra" is commonly used in healing work. It is taken from the Eastern Ayurvedic system of healing. Other old esoteric teachings in various traditions teach the same idea that the body has seven major and distinct zones of energy, including the Jewish Kaballah and the seven Catholic sacraments. For more information on these ideas, refer to the work of Caroline Myss. Chakras are the primary energy centers along the human spinal column from the first at the spinal, base (root) chakra to the seventh chakra at the top of the head (crown). Each of these areas is associated with a specific gland, which may help the Western mind understand chakra associations to body areas and what the potential impact on the body may be. These glands are indicated below as well as common associations made to areas of our lives thought to be governed by these energy centers. Each center is associated to a color, which corresponds to the rainbow or prism of light.

	Glands	*Principal Aspect*
1st	Root - adrenals	Grounding, survival, material world, tribal law and connections
2nd	Navel - reproductive	Sexuality, desires, work, creativity, relationships, energy output
3rd	Solar plexus - spleen	Ego, personal power, inspiration for life, gut-level intuition
4th	Heart - thymus	Love, emotions, compassion, conscious heart, forgiveness
5th	Throat - thyroid	Communication, self-expression, will, manifestation
6th	Brow (third eye) - pituitary	Mind, intuition, visions,
7th	Crown - pineal	Spirituality

Chakra/Mineral Cross Reference

First Chakra: *Base of Spine, Root* ***Colors:*** *Red, Black, Brown* ***Polarity:*** *Neutral*

Acanthite
Agate - Bull's Eye
Agate - Dry Head
Agate - Fire (also 6th)
Agate - Pine Moss
Agate - Plasma
Agate - Plume (also 2nd)
Agate - Turritella
Agate - Zebra
Alabaster - Orange/Peach
Allanite
Amethyst - Thunder Bay
Anasazi Red Ocher
Andalusite (to 4th)
Apache Gold (and upper)
Apache Tear - Obsidian
Aragonite - Brown
Arfvedsonite
Arsenopyrite
Astrophyllite
Augite
Aventurine - Red
Axinite
Beryl - Common, Red
Bixbyite (to upper)
Bloodstone
Bronzite
Brookite
Buddstone
Bustamite
Calcite - Blue (to 7th)
Cassiterite
Chalcedony - Black
Chalcopyrite (to 7th)
Chiastolite
Childrenite (to 3rd)
Cinnabar (and 12th)
Clinoclase (to 6th)
Coal
Colorado Sunset
Coprolite
Coral - Red, Pink, White
Crotia Diamond
Datolite Nodule
Descloizite
Diopside
Eilat Stone
Enargite
Eosphorite
Erythrite
Eudialyte (to 4th)
Eudialyte with Riebeckite
Fluorite with Pyrite
Fossils
Garnet - Andradite
Garnet - Melanite (Black)
Garnet - Red
Gehlinite
Gibbsite
Glendonite
Goethite
Granite
Groutite
Hedenbergite (to 5th)
Herderite
Heubnerite
Hickoryite - Jasper (to 3rd)
Hornblende (to 7th)
Iron
Jade - Black
Jasper - Brown
Jasper - Orbicular
Jasper - "Pigeon Blood"
Jasper - Sweet Home
Jasper - Willow Creek
Jet
Kammererite
Lanarite
Laurel Leaf Fossil
Lead
Lepidocrocite
Limonite
Lodestone
Magnesite
Manganotantalite
Margarite
Marmatite
Mercury (to 6th)
Metals
Mookaite
Mottramite
Mozarkite
Neptunite
Nipigon Stone
Ocean Picture Stone - with brown
Obsidian - Black
Obsidian - Fire
Obsidian - Mahogany
Obsidian - Snowflake
Onyx - Black
Opal - Cherry
Phenakite (and 7th)
Pipestone (and 4th)
Platinum (and 8th)
Plattnerite
Prase (4th and 7th)
Proustite
Psilomelane
Pyrargyrite
Pyrite
Pyrophyllite
Pyroxmangite
Quartz - Midnight Black
Quartz - Prasio
Quartz - Red
Quartz - Strawberry
Realgar
Realgar/Orpiment
Riebeckite (and 2nd)
Rhodazite
Rhodonite
Romanechite
Ruby (to 4th)
Rutile - Red
Sagebrush - Petrified
listed under Chalcedony
Schorl - Tourmaline (Black)
Selenite - Black Satin
Septarian Nodule - Red
Silicon (and 7th)
Smoky Quartz
Sphalerite
Spinel - Red, Ruby, Black
Staurolite
Tibetite
Tiger Iron
Titanium on Quartz
Topaz - Brown (and 2nd)
Topazolite
Touchstone
Turgite
Turtle Back Stone
Vanadinite
Variscite (to 4th)
Vesuvianite - Black
Wad
Whiteite
Wulfenite
Zebra Stone
Zoisite with Ruby
Zincite - Red

Second Chakra: *Navel, Abdomen* *Color:* *Orange* *Polarity:* *Feminine*

Alabaster - Orange/Peach
Alexandrite (7th and 8th)
Amber
Aragonite - Brown
Aventurine - Peach
Bronzite
Bustamite
Calcite - Orange
Carnelian
Chalcopyrite
Citrine - Orange
Colorado Sunset
Crocoite
Epidote (to 4th)
Galena
Garnet - Andradite
Gold (to 4th)
Gypsum
Halite (and 3rd)
Hanksite (also 3rd and 9th)
Hematite - Rainbow
Heliodor - Gold
Jade - Brown
Jasper - Brecciated
Jasper - Bruneau
Jasper - Rainforest
Jasper - Spider Web
Obsidian - Fire
Onyx - Honey (and 3rd)
Opal - Cherry
Petersite
Phacolite
Pyromorphite - Golden
Pyrrhotite
Quartz - Prasio
Quartz - Tibetan
Rhodochrosite (to 4th)
Riebeckite
Sagebrush - Petrified listed under Chalcedony
Sard
Septarian Nodule - Yellow
Serandite
Scheelite
Stilbite - Orange
Sturmanite
Titanium on Quartz
Topaz - Imperial
Tourmaline - Uvite (Green)
Trinitite
Vanadinite
Vesuvianite - Brown
Wad
Whiteite
Wulfenite
Zincite (and 3rd)

Third Chakra: *Solar Plexus* *Color:* *Yellow* *Polarity:* *Masculine*

Adamite - Yellow
Alabaster - Orange/Peach
Amber
Anglesite (and higher)
Annabergite (and 4th)
Apatite - Yellow. Golden
Aragonite - Golden
Autunite
Brazilianite
Calcite - Iceland Spar
Calcite - White, Honey, Green
Chalcanthite
Chalcocite
Chalcopyrite
Childrenite
Chrysoberyl
Chrysoprase - Lemon, Dendritic
Chrysotile
Citrine - Yellow
Crandallite and Variscite
Darwin Glass
Dravite - Tourmaline (Brown)
Entigrite
Fluorite - Yellow
Galena (and 4th)
Garnet - Andradite
Garnet - Uvarovite
Garnet - Yellow
Gaspeite
Heliodor - Golden Yellow
Hematite - Rainbow
Jasper - Bruneau
Jasper - Butterfly
Jasper - Imperial Green
Legrandite
Lepidolite - Yellow
Lizard Stone
Manganite
Manganocalcite
Mimetite
Nipigon Stone
Opal - Black
Opal - Cherry
Orpiment
Pyrrhotite
Quartz - Golden Healer
Sapphire - Yellow
Scheelite
Septarian Nodule
Siderite
Soapstone (to 6th)
Sphaerocobaltite
Sturmanite
Sulfur
Sunstone - Heliolite (and 4th)
Tektite
Titanium on Quartz
Topaz - Orange, Imperial
Wavellite
Witherite
Vanadinite
Zincite - Yellow

Fourth Chakra: *Heart* *Colors:* *Green, Pink* *Polarity:* *Feminine*

Actinolite
Adamite Green-Cuprian
Adularia (and all skin)
Agate - Angelskin
Agate - Evergreen (to 7th)
Agate - Moss
Agate - Plasma
Amethyst - Chlorite-included
Annabergite
Apatite - Blue/Green
Apatite - Green
Atacamite
Autunite
Aventurine - Green
Barbertonite
Bloodstone
Brazilianite
Brochantite
Buddstone
Burnite
Calcite - Green, Pink
Celadonite
Celadonite in Quartz
Cerussite and Malachite
Chalcedony Rose
Chalcopyrite
Chromite and Malachite
Chrysocolla
Chrysoprase (and 5th)
Conichalcite
Coral - Pink, Red
Crandallite and Variscite
Cuprite
Danburite - Pink
Darwin Glass
Dioptase
Dioptase and Kinoite (to 6th)
Edenite in Quartz
Eilat Stone (to 6th)
Emerald
Eudialyte with Riebeckite
Fluorite - Rainbow (to 6th)
Forsterite
Fuchsite
Garnet - Hydrogrossular (Green)
Garnet - Hydrogrossular (Pink)
Garnet - Uvarovite
Gem Silica (5th and 6th)
Gold (to 2nd)
Granite
Hedenbergite
Heliodor - Lime Green
Hematite - Rainbow
Hiddenite - Green, Pink
Hornblende (to 1st)
Jade - Pink, Red
Kunzite (and up)
Kyanite - Green
Lapis Nevada
Lizard Stone
Malachite
Manganite
Manganocalcite
Manganotantalite
Mica - Pink
Moonstone - Green, Peach, White
Morganite (to upper)
Mtorolite
Nambulite
Nunderite
Obsidian - Pink Snowflake
Obsidianite - Emerald Green
Ocean Picture Stone
Onyx - Green, Honey
Opal
Pecos Diamond
Peridot
Prase - Green
Pyrite
Pyrrhotite
Pyroxmangite
Quartz - Blue (and 5th)
Quartz - Chlorite-included (and 7th)
Quartz - Orchid, Rose
Rhodochrosite (to 2nd)
Rhyolite (also 3rd)
Rosasite (to 6th)
Sapphire - Green
Scapolite - Pink
Serpentine
Smithsonite - Pink
Sphaerocobaltite
Spinel - Green, Pink
Stichtite (to crown)
Thulite
Titanium on Quartz
Topaz - Champagne
Tourmaline - Bi-color
Tourmaline - Green
Tourmaline - Pink
Tourmaline - Rubellite
Tremolite
Turquoise - Green
Turtle Back Stone
Unakite
Variscite (to 1st)
Verdite (to 6th)
Zebra Stone
Zincite - Green

Fifth Chakra: Throat   *Color:* Light Blue   *Polarity:* Masculine

Acanthite
Agate - Blue Lace (to 6th)
Agate - Ellensburg Blue (to 6th)
Agate - Evergreen
Amazonite (Microcline)
Amblygonite (through 7th)
Angelite
Anhydrite
Apatite - Blue and Blue/Green
Apatite - Blue on Clevelandite
Apatite - Purple
Aquamarine (Blue Beryl)
Aragonite - Laurium
Azurite/Malachite
Boji Stone™ - Pop Rock
Brucite
Celestite (toward 7th)
Chalcedony - Blue
Chrysocolla (also 6th)
Chrysolite
Fluorite - Blue
Garnet - Andradite
Garnet - Grossularite (Orange)
Gem Silica
Garnet - Grossular (Pink/Green)
Hematite - Rainbow
Heulandite
Inesite
Iolite (to 9th)
Jade - Blue
Jade - Green
Jadeite
Jasper - Imperial Red
Jasper - Juniper Blue
Kidwellite
Kinoite
Lapis Lace
Larimar
Libyan Blue
Mesolite (and 7th)
Morrisonite - Jasper (and 6th)
Obsidianite - Light Blue
Ocean Picture Stone
Onyx - Light
Opal - Blue
Petersite
Pyrite
Quartz - Blue (to 6th)
Quartz - Siberian Blue
Sardonyx
Smithsonite - Blue (and 6th)
Spinel - Blue (and 6th)
Strombolite
Sugilite (and 6th)
Tiger Eye - Cherry
Tiger Eye - Yellow
Tin
Titanium on Quartz
Trinitite
Tunnelite
Turquoise - Blue
Verdite
Youngite (and 6th)
Zincite - Blue/Gray

Sixth Chakra: *Third Eye, Brow, Forehead* **Colors:** *Blue, Dark Blue, Indigo*
Polarity: *Feminine*

Agate - Angelskin
Agate - Blue Lace
Agate - Ellensburg Blue
Agate - Glory Blue
Agate - Holly Blue
Ajoite
Amethyst - Brazil
Amethyst - Central Mexico Phantom
Amethyst - Chlorite-included
Amethyst - Chevron
Amethyst - Vera Cruz, Mexico
Amethyst - Siberian Blue
Amethyst - Thunder Bay, Ontario
Amethyst - United States
Amethyst - Uruguay
Apatite - Blue
Apatite - Blue on Clevelandite
Apatite - Purple
Aurichalcite
Aventurine - Blue, Lavender
Azurite
Barbertonite
Barite - Blue
Benitoite
Bertrandite
Beryllium and Fluorite
Bismuth
Boleite
Burnite
Calcite - Blue
Carletonite (to 7th)
Cavansite
Ceruleite
Chalcanthite
Chalcedony - Blue
Chalcocite
Charoite
Cookeite
Coral - Red Horn
Covellite
Creedite (to 7th)
Dumortierite - Blue. Red
Euclase
Fluorite - Blue, Purple
Garnet - Hydrogrossular (Pink, Green)
Garnierite
Halite - Blue
Hawk's Eye - Blue Tiger Eye
Hematite (and 7th)
Hematite - Rainbow
Hemimorphite (to 7th)
Heubnerite (to 7th)
Heulandite
Iolite
Jade - Blue, Lavender
Jadeite - Lilac
Kammererite
Kidwellite
Kottigite
Kyanite
Labradorite
Laurel Leaf Fossil
Lapis Lazuli
Lazurite
Lepidolite
Libyan Blue
Linarite
Magnesite in Quartz
Margarite
Meteorite (also 5th)
Moldavite Tektite
Morrisonite - Jasper (to 7th)
Mother of Pearl
Muscovite (to 7th)
Nipigon Stone
Opal - Black, Blue
Oyster Shell
Papagoite-included Quartz
Paua Shell
Pearl
Peristerite
Petosky Stone
Purpurite (to 7th)
Quahog
Quartz - Blue Rutile
Quartz - Lavender
Quartz - Siberian Blue
Quartz - Silver Rainbow
Richterite
Royalite
Sagenite
Sapphire - Blue, Purple
Scapolite - Purple
Sodalite
Shattuckite
Stewartsonite
Strombolite
Sugilite - Luvulite, Royal Azel
Sugilite - Blue
Tanzanite
Tourmaline - Blue
Tremolite - Pink
Uraninite (and 7th)
Verdite
Vesuvianite - Lavender
Violane
Vivianite
Wavellite
Wolframite
Wulfenite

Seventh Chakra: *Crown, Top of Head* *Colors:* *Violet, White* *Polarity:* *Neutral*

Adamite (Purple)
Agate - Burro Creek
Agate - Dendritic
Agate - Evergreen
Agate - Montana Picture
Agate - Plasma
Agate - Polka Dot
Agate - Zebra
Ajoite
Alabaster
Allanite
Amethyst - Central Mexico
Amethyst - Chevron
Amethyst - Chlorite-Included
Amethyst - Phantom
Analcime
Apatite - Blue on Clevelandite
Aragonite - White
Artinite
Astrophyllite
Aventurine - Lavender
Azeztulite
Barite - Clear (also 12th)
Barite - Desert Rose
Barthite
Bertrandite
Beryl - Golden
Beryllium and Fluorite
Buddstone
Celadonite in Quartz
Chabazite
Colemanite
Cookeite
Coral - Red Horn
Crotia Diamond
Cubic Zirconia
Danburite - Clear, Milky, Pink
Diamond
Dumortierite - Red
Edenite in Quartz
Eudialyte with Riebeckite
Feldspar
Geodes - Quartz
Glauberite
Goshenite
Garnet - Grossular (Pink/Green)
Granite
Hematite - Rainbow
Herkimer Diamond (and 6th)
Howlite
Ilvaite
Jade - Lavender, Lilac
Jasper - Biggs
Jasper - Imperial Green
Jasper - Leopardskin
Jasper - "Pigeon Blood"
Jasper - Rainforest
Jasper - Snakeskin
Kernite
Lepidolite
Magnesite in Quartz
Mesolite
Microcline
Natrolite
Oaxaca Diamond
Obsidian - Pink Snowflake
Ocean Picture Stone
Okenite
Onyx - White
Opal - Blue, Hyalite, White
Papagoite-included Quartz
Petalite
Phenakite
Prase - Green
Pyrite
Pyrolucite
Pyrophyllite
Pyroxmangite
Quartz - Blue Rutile
Quartz - Castle Tip
Quartz - Celestial
Quartz - Dragon's Tooth
Quartz - Diamontina (Star Seed)
Quartz - Drusy
Quartz - Hematite-included
Quartz - Phantom
Quartz - Pleiadean
Quartz - Silver Rainbow
Quartz - Solution
Quartz - Tangerine
Quartzite
Royalite
Sapphire (and 9th)
Scapolite - Purple
Scolecite
Selenite - Black Satin
Selenite - White/Clear
Sillimanite
Spinel - White
Sugilite - Blue, Purple
Tarnowitzite
Titanium on Quartz
Topaz - Golden
Topaz - White
Tremolite - White
Tunnelite
Vesuvianite - Lavender
Wardite
Witherite
Wollastonite
Zircon (and 6th)

Upper Chakra Activation

There are many other energy centers (chakras) in the physical and etheric bodies, but five important ones exist above the seventh (crown). A brief discussion of these follows:

The **eighth chakra** coordinates the etheric and emotional bodies and relates to our global awareness

The **ninth chakra** coordinates the emotional with the mental body and relates us to our solar system

The **tenth chakra** relates the mental to the astral body and relates us to the Milky Way galaxy

The **eleventh chakra** integrates the astral with the causal body and brings in universal energies

The **twelfth chakra** integrates the causal with all the spiritual bodies and links us to the undifferentiated whole

These are energetic chakras in the auric and astral field of each individual. The employment of crystals for developing channeling of energy and strengthening the conduit of the eighth through the twelfth chakras is a critical matter not left to children as a plaything. Incorporating vibratory shift into the body based on spiritual practice puts incredible stress on the nervous system. Attempting physical conversion based on exposure to mineral frequencies requires conscious attention and sobriety. One must be solemnly ready to accept the responsibilities of power and the toll energy work takes on the nervous system. Traditions which affect consciousness evolution have guarded hidden knowledge, because many in the world were simply not ready, willing or able to maintain the vibratory shift without losing themselves in the process. One must not assume that interest in the metaphysical ensures that one is meant to do the work of the adept. Some bodies are more ready than others through genetic inheritance. Some souls have more training and experience from other realms (to remember) available to them. Creative and mental genius is a strong indicator that kundalini flow is already occurring at a significant rate in an individual. For any individual, increasing this flow is a solemn undertaking. For more information on kundalini, refer to the book *Living with Kundalini* by Gopa Krishna. While all humans transmit and receive the life force, light and information of the universe, no two people stand in the same wavelength.

Minerals for the five chakras above the crown chakra to assist with healing and psychic work can be reversed in order based on gender. Difficult weather conditions, traumatic human events, and earthquakes and aftershocks, etc. may require healers and psychics to use minerals which enhance or optimize energy frequencies in order to do their work. When the weather is overcast or an inversion has been in the area, recent human events involved trauma or the day is believed to be negative (such as Friday the 13th), humans may need frequency optimizers to increase signal strength and reliability, to reduce outside "noise and static," and to reduce drain on or damage to the nervous system. Gender affects vibration and frequency entering the human form through refraction differences caused by bone density, hormones, muscle mass to fat ratio, and menstruation. For women, the order of frequency optimizers listed below is reversed.

The minerals are listed in the order of which is most useful to obtain first. Having one of each type is ideal. Men can best use frequency optimizers in this order: 1) **optical calcite** (pink or salmon tone is best); 2) **phantom quartz** with phantom interstices resonating to the birth information of the individual, or other characteristics, such as health needs; 3) **fluorite**; 4) **blue/green/black tourmaline**; and 5) **spheres**. The area of spheres is the most interchangeable, and the one most sensitive to gender. Spheres resonate to the feminine principle, therefore more women will naturally resonate to their energies at first. Spheres affect the primary psychic energy translation conduit in the body (third eye) where information begins to take on its associative links and power familiar to each individual through experience. Frequency optimizer spheres can be made of any form of obsidian (especially snowflake), quartz (especially rose), beryl, and chalcedony. The resonance found in obsidian concentrates the descent of light so that it can penetrate surrounding influences. Obsidian emits a sound frequency which provides protection and enhances intuition.

A broad range of other frequency optimizers are available to people. Water is an excellent frequency enhancer. Its tonality is not well defined, so it takes practice to utilize water for consciousness, healing and psychic work. After familiarization or because of an individual's resonance to water, the sacred potency of water cannot be matched. Notice the ability of water to enhance intuition while bathing or washing dishes. Fountains are excellent for all healing and psychic work areas. Sacred sand is used in ritual which will be understood by Indian medicine traditionalists. Refer to references on the ephemerality of sand painting for more information. Sand is the raw form of silica, which is an excellent medium for passing light frequency across a spectrum and receiving insight from spirit. The act of locating one's sand need not be any more formal than selecting an important spot in one's life where one felt certain high vibrations, and gathering a bit of sand to be part of one's medicine bundle. Flame and smoke are spirit frequency enhancers. Flame and smoke provide concrete power, yet ephemeral form. One cannot capture them without destroying their essence. The lesson here for practitioners is that connection to the higher self, the collective unconscious, Source cannot be captured without changing irrevocably the vital essence being sought. Science has shown us that we can know only one thing at a time about electrons, either its position or its speed, but not both. Being a conduit of energy requires being willing to let go of control. One is not a vessel which is filled and sealed with perfect knowledge contained within forever. One is merely like a hologram box where the vibrations "enter" to take form for moments so that the learning can be spoken of or recorded. But the vital essence passes through as if the individual were merely a tube with open ends.

Stones and minerals can be worked with in many elaborate and over-blown ways, but this overemphasis on hyper-connection is necessary only for those who have trouble making the connection at all or who have more than an ounce of drama to express as their healing style. For people already manifesting psychic ability, creative and mental genius, or spiritual insight, stones can be used during sleep and meditation with significant benefits. Attunement of frequencies during sleep teaches humans "the ropes" of energy as a natural, inevitable skill, and reduces the tendency to treat minerals as devices which make consciousness shift and evolution happen when they are "turned on." In using stones and minerals, it is best for most people to maintain natural interaction. Mineral super-technology requires delicate balance to control, and it can create severe disharmony as well as unwanted negative results.

Upper Chakras - 8th through 12th above Crown

8th Chakra Color/Polarity: Magenta/Feminine
9th Chakra Color/Polarity: Purple/Masculine
10th Chakra Color/Polarity: Copper/Feminine
11th Chakra Color/Polarity: Gold/Masculine
12th Chakra Color/Polarity: Silver/Feminine

Agate - Burro Creek
Apophyllite
Azeztulite
Barite - Gold
Calcite - Peach
Cavansite
Cerussite
Chalcedony - Black
Colemanite
Danburite
Datolite
Gmelinite
Heliodor - Golden
Hematite - Rainbow
Howlite
Kinoite
Mica - Gold, Silver
Moonstone - Rainbow
Petersite
Prehnite
Purpurite
Quartz - Tibetan
Scapolite - Purple
Serafinite
Silver
Stibnite
Stilbite
Titanium on Quartz
Torbernite
Tourmaline - Blue
Trinitite
Tunnelite

Auric Chakras

Betafite
Garnet - Rhodalite
Hematite - Rainbow
Marcasite
Moonstone - Rainbow
Obsidian - Black
Orthoclase
Pink Limbcast
Quartz - "Aqua Aura"™
Titanium on Quartz
Turgite
Zektzerite

Etheric Chakras

Agate - Polyhedron
Betafite
Halite - Pink
Mariposite
Obsidianite - Emerald Green
Ocean Picture Stone
Rhodochrosite
Quartz - Golden Healer

All Chakras

Abalone Shell
Albite
Anapaite
Aragonite - Purple
Agate - Botswana
Biotite Mica
Brass
Bronze
Cacoxenite
Calcite - Honey, White
Chert
Clinochlore
Colemanite
Copper
Datolite
Fluorichterite
Fluorite
Garnet - Demantoid
Gold
Halite
Hematite - Rainbow
Jade (primarily 4th through 6th)
Jamesonite
Lazulite (especially 6th)
Lingam
Lithium
Marcasite and Galena
Meteorite
Moonstone - Clear
Obsidian - Rainbow, Rootbeer
Obsidian - Snowflake
Olivine
Pearl
Quartz - "Aqua Aura"™
Quartz - Fisher Mountain Blue Phantom
Quartz - Iron Sulfide-included
Quartz - Rutilated
Silver
Spectralite
Sphene
Stibnite
Sustamite
Tourmaline
Ulexite
Vesuvianite
Willemite
Zircon

Other Influences

Agate - Iris - hands
Agate - Snakeskin - palms and soles of feet
Azulicite - hands and feet
Flint, Chert - varies with color
Glass - varies with color
Hiddenite - Green
Jasper - varies with color
Kunzite - Pink
Marble - varies with color
Porphyry, Flowerstone - varies with color
Schist - fingers and meridians
Spodumene - varies with color

APPENDIX 2

A Spiritual Apothecary

The most consistent healing tools are minerals because of their organized structure, semi-conducting abilities, and color characteristics. Fewer allergic reactions are likely to occur with minerals and stages of healing are deep and slow. Minerals normally do not address traumatic healing situations the way herbs or some animals do. Minerals affect change through inexorable vibratory influence. We suggest working with a mineral for as continuously as possible for thirty days before reaching a conclusion about its affect. This increases personal synchronization with mineral frequencies.

Healers who use minerals can benefit from studying the characteristics of the Spiritual Apothecary minerals listed below. Developing an awareness of how these minerals affect the subtle and physical bodies improves the efficiency of subtle energy healing processes. **First**, the **auric** and **etheric bodies** are **cleansed** and then **opened** by placing minerals in those categories near or on the body. **Second**, specific **healing** minerals are placed on or near the physical body. **Third**, the **auric** and **etheric bodies** are **conditioned** and **sealed**.

Ideally, the cleansing phase cleans the slate of the individual receiving energy. The opener minerals provide safety, discernment, and intense infusions of light. The healing phase contains a broad variety of choices and energetic influences. Conditioner minerals balance, integrate, and harmonize the subtle and physical bodies. All sealer minerals in one way or another indicate to the individual that closure is occurring and all energy centers receive protection energy.

When this concept was first suggested to us during psychic readings, the bare bones of the list was all that was given. Studying this skeleton over a period of years resulted in noticing a pattern involving schorl (black tourmaline) and the beryl family of minerals.

The single most important family of minerals listed in the Spiritual Apothecary is beryl, as all known members of this family are represented: **Aquamarine, Goshenite, Heliodor, Morganite, Emerald,** and **Red Beryl**. Four other important minerals need to be noted: **Schorl, Danburite, Herkimer Diamond** and **Chlorite-included Quartz**. Schorl is listed in five areas: auric cleansers, etheric cleansers, healers under all tourmalines, auric sealers and etheric sealers, and theoretically, could replace emerald and red beryl in the etheric sealers, if suitably priced examples cannot be obtained. With seven minerals (aquamarine, goshenite, heliodor, morganite, schorl, Herkimer diamond and chlorite-included quartz), you can accomplish all nine steps to address the energy systems of the physical and subtle bodies. By adding emerald and danburite, your essential apothecary can be comprised of nine minerals, which is a powerful number to work with, because it is karma-free.

All nine minerals can be placed around the body simultaneously or they can be introduced in successive order. The chlorite-included quartz and danburite can be utilized by the practitioner for specific energy movements.

The total number of minerals in the Spiritual Apothecary is 108 — the Buddha number. Be sure to note that Barite, under Auric Conditioners, includes both clear and blue varieties.

Let your intuition and budget help you select your personal apothecary of healing stones in each category. Individual frequencies may demand different mineral choices altogether than those listed above or below. The minerals in each category below provide unique signature spins on a common theme. You can choose to work with deep blue apatite to select minerals for your own spiritual apothecary.

Cleansers and Openers

AURIC CLEANSERS

Apophyllite
Cerussite
Chrysoprase
Clinozoisite
Dioptase
Feldspar
Heliodor
Howlite
Jade
Kyanite
Legrandite
Rhodochrosite
Schorl
Sunstone

ETHERIC CLEANSERS

Amber
Calcite (Clear, Yellow)
Dravite
Feldspar
Fire agate
Fluorite (Golden, Yellow)
Garnet (Uvarovite)
Halite
Larimar
Lazulite
Malachite
Schorl
Sulfur

AURIC OPENERS

Apatite (Yellow)
Chert/Flint
Danburite
Dumortierite
Herkimer Diamond
Iolite
Selenite
Sunstone
Tiger Eye (Blue Hawk's Eye)

ETHERIC OPENERS

Azurite (Azurite/Malachite)
Calcite (Blue)
Chert/Flint
Fluorite (Purple)
Gem Silica (Chrysocolla)
Girasol Opal
Goshenite
Ocean Picture Stone (copper silicate)
Phenakite

HEALERS

Anhydrite
Alabaster
Apatite (Blue)
Boleite
Cassiterite
Chalcedony
Charoite
Chrysoberyl
Fluorichterite
Garnet (All)
Gold
Halite
Jade (All)
Kunzite
Lepidolite
Morganite
Obsidian
Purpurite
Quartz
Sardonyx
Schist
Serandite
Serpentine
Silver
Stilbite
Sugilite
Topaz
Tourmaline (All)
Turquoise
Variscite
Vesuvianite (All)
Vivianite
Willemite

Conditioners and Sealers

AURIC CONDITIONERS

Amethyst
Aquamarine
Barite (Clear and Blue)
Chalcopyrite
Copper
Holly Blue Agate
Labradorite
Lithium
Moonstone (Rainbow)
Rhodochrosite

ETHERIC CONDITIONERS

Aurichalcite
Botswana Agate
Chlorite-included Quartz
Cinnabar
Crocoite
Garnet (Red)
Mother-of-Pearl
Opal
Sapphire
Wulfenite
Zoisite

AURIC SEALERS

Celestite
Chalcedony (especially Black)
Coral
Euclase
Gold
Muscovite
Obsidian (especially Apache Tear)
Schorl
Tanzanite
Tektite (Moldavite)

ETHERIC SEALERS

Aragonite
Beryl (Red)
Chalcedony (Black)
Citrine
Emerald
Howlite
Lapis Lazuli
Moonstone (Clear/White)
Ruby
Schorl
Unakite

References

Health

The Natural Health Book; Dorothy Hall; Charles Scribner's Sons, 1976.

Mineralogy

Chambers's Mineralogical Dictionary; Chemical Publishing Co., Inc., 1948.

The Collector's Encyclopedia of Rocks and Minerals; Edited by A.F.L. Deeson; Clarkson N. Potter, Inc., 1973.

Encyclopedia of Minerals; Willard L. Roberts, Thomas J. Campbell, George R. Rapp, Jr.; Van Nostrand Reinhold Company, New York; 1990.

Fleischer's Glossary of Mineral Species; Joseph A. Mandarino, The Mineralogical Record, Inc., Tucson, 1999.

Gems, Crystals & Minerals; Anna S. Sofianides and George E. Harlow, the American Museum of History; Simon and Schuster, 1990.

Glossary of Mineral Species; by Michael Fleischer; The Mineralogical Record, Inc., Tucson; 1983, 1987.

Simon and Schuster's Guide to Rocks and Minerals; edited by Martin Prinz, George Harlow and Joseph Peters, The American Museum of History; Arnoldo Mondadori Editore S.p.A., Milan; 1977, 1978.

The World of Fossils; Giovanni Pinna; Bounty Books; 1973.

Symbology

An Illustrated Encyclopedia of Traditional Symbols; J.C. Cooper; Thames and Hudson, Ltd., London, 1978.

Dictionary of Symbols; Jean Chevalier and Alain Gheerbrant, Translated by John Buchanan-Brown; Penguin Books, 1996.

Medicine Cards: The Discovery of Power through the Ways of Animals; Jamie Sams and David Carson; Bear and Company, Santa Fe, New Mexico; 1988

INDEX

Healing and Integration Properties to Minerals

A

B

C

D

E

G

H

J

M

N

O

P

S

U

W

Y

Z